# GCSE IN A WEEK

## Chemistry

**Emma Poole**

# Revision Planner

# How Science Works

## Ideas

Sometimes our opinions are based on our own prejudices, i.e. what we personally like or dislike.

At other times, our opinions can be based on scientific evidence. These opinions are based on reliable and valid evidence that can be used to back up our opinion.

## Variables

- An **independent** variable is the variable that we choose to change to see what happens.
- A **dependent** variable is the variable that we measure.
- A **continuous** variable, e.g. time or mass, can have any numerical value.
- An **ordered** variable, e.g. small, medium or large, can be listed in order.
- A **discrete** variable can have any value that is a whole number, e.g. 1, 2.
- A **categoric** variable is a variable that can be labelled, e.g. red, blue.
- We use **line graphs** to present data where the independent variable and the dependent variable are both continuous. A line of best fit can be used to show the relationship between variables.
- **Bar graphs** are used to present data when the independent variable is categoric and the dependent variable is continuous.

## Evidence

Evidence should be:

- **reliable** and **repeatable** (if you do it again you get the same result)
- **accurate** (close to the true value).

Scientists often try to find links between variables.

Links can be:

- causal – a change in one variable produces a change in the other variable
- a chance occurrence
- due to an association, where both of the observed variables are linked by a third variable.

We can use our existing models and ideas to suggest why something happens. This is called a **hypothesis**. We can use this hypothesis to make a **prediction** that can be tested.

When data is collected, if it does not back up our original models and ideas, we need to check that the data is valid. If it is valid we need to go back and change our original models and ideas.

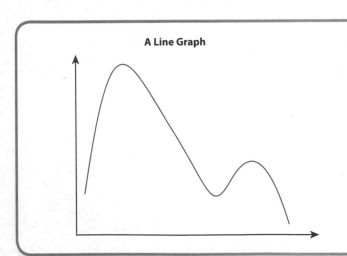

**A Line Graph**

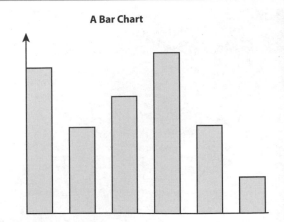

**A Bar Chart**

# Science in Society

Sometimes scientists investigate subjects that have social consequences, e.g. food safety. When this happens, decisions may be based on a combination of the evidence and other factors, such as bias or political considerations.

Although science is helping us to understand more about our world there are still some questions that we cannot answer, such as: 'Is there life on other planets?' Some questions are for everyone in society to answer, not just scientists, such as: 'Should we clone humans?'

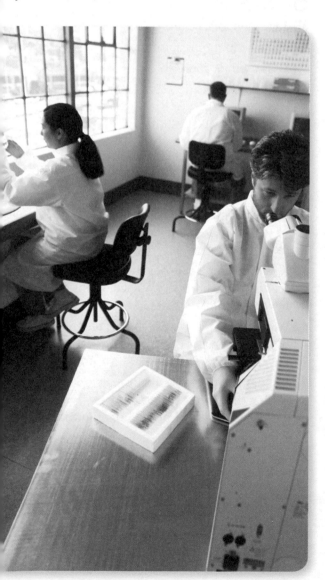

## SUMMARY

- There are different types of variable.
- Scientists try to find links between variables.
- Some questions cannot be answered by scientists.

## QUESTIONS

### QUICK TEST

1. What is an independent variable?

2. What is a dependent variable?

3. What is an ordered variable?

4. What is a discrete variable?

5. What does accurate mean?

6. When should a line graph be used to present the results of an investigation?

### EXAM PRACTICE

1. A student carries out an experiment to find out how the force applied to a spring affects the length of the spring.

    a) What is the independent variable? **(1 mark)**

    b) What is the dependent variable? **(1 mark)**

    c) Suggest a variable that must be controlled to make it a fair test. **(1 mark)**

# Atomic Structure

## Atoms

Elements are made of only one type of atom. Atoms consist of a small, central **nucleus** (that contains protons and neutrons) surrounded by shells of electrons.

All atoms of the same element have the same number of protons.

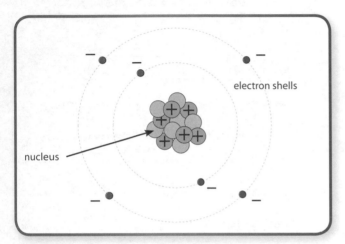

electron shells

nucleus

Protons have a mass of 1 and a charge of $1^+$. Electrons have a negligible mass and a charge of $1^-$. Neutrons have a mass of 1 and no charge.

## Bonding

Atoms can join together by:

- sharing pairs of electrons (this is called a **covalent** bond)
- transferring electrons (this forms ions – the attraction between oppositely charged ions is called an **ionic** bond).

## The Periodic Table

- There are about 100 different elements, which are displayed in the **periodic table**.
- The horizontal rows are called **periods**.
- The vertical columns are called **groups**.
- Group 1 of the periodic table is also known as the alkali metals.
- Group 7 is known as the halogens.
- Group 0 is known as the noble gases.

When the periodic table was first designed, many of the elements which we know today had yet to be discovered. Gaps were left in the table and detailed predictions were made about what the new elements would be like.

## Symbols

Elements can be represented using symbols.

Symbols are one- or two-letter codes. For example:

- O is used for oxygen
- C is used for carbon
- Ca is used for calcium
- K is used for potassium
- Na is used for sodium
- Pb is used for lead
- Fe is used for iron.

## Formulae

The **chemical formulae** tells us the number and type of atoms present. For example:

- An oxygen molecule has the formula $O_2$. This tells us that an oxygen molecule consists of two oxygen atoms.
- A nitrogen molecule has the formula $N_2$. This tells us that a nitrogen molecule consists of two nitrogen atoms.
- Water has the formula $H_2O$. This tells us that a water molecule consists of two hydrogen atoms and one oxygen atom.
- Copper(II) sulfate has the formula $CuSO_4$. This tells us that the ratio of the atoms is:
  – one copper atom : one sulfur atom : four oxygen atoms.
- Calcium hydroxide has the formula $Ca(OH)_2$. This tells us that the ratio of the atoms is:
  – one calcium atom : two oxygen atoms : two hydrogen atoms.
- Hydrated means that the compound contains water. For example, the formula for hydrated copper(II) sulfate is $CuSO_4.5H_2O$

# Equations

We can use word equations to sum up what happens during a chemical reaction.

For example, when carbon is burned in plenty of oxygen, carbon dioxide is produced. This can be written as:

carbon + oxygen → carbon dioxide

This reaction can also be summed up using the symbol equation:

$$C + O_2 \rightarrow CO_2$$

When hydrogen is burnt in oxygen, water vapour is produced. This can be written as:

hydrogen + oxygen → water vapour

This can also be written as:

$$2H_2 + O_2 \rightarrow 2H_2O$$

During a chemical reaction, atoms are not created or destroyed, they are simply rearranged. This means that there must be equal numbers of each type of atom on both sides of the equation.

## SUMMARY

- Atoms contain protons, neutrons and electrons.
- Elements can be represented by symbols and displayed in the periodic table.
- Formulae can be used to show the type and number of atoms present in a substance.
- Word and symbol equations can be used to sum up chemical reactions.

## QUESTIONS

### QUICK TEST

1. What do all the atoms of the same element have in common?

2. What is the small, central part of an atom called?

3. What are found in shells around the central part of an atom?

4. Where are the names and symbols of elements displayed?

5. Roughly how many elements are there?

### EXAM PRACTICE

1. Calcium carbonate ($CaCO_3$) reacts with hydrochloric acid to produce the salt calcium chloride ($CaCl_2$), water ($H_2O$) and carbon dioxide ($CO_2$).

   a) Write a word equation for the reaction.
   **(1 mark)**

   b) Write down the type and number of atoms present in one molecule of carbon dioxide.
   **(1 mark)**

# The Periodic Table

## History

In the modern periodic table there are about 100 different elements.

The horizontal rows are called **periods**, and the vertical columns are called **groups**.

Elements with similar properties occur periodically, so it is called the periodic table.

As the elements were discovered, scientists tried to place them into a logical order. John Newlands and Dmitri Mendeleev tried to place the elements in order of increasing atomic weight.

Mendeleev realised that, although this was a good basis, if he stuck to this order too rigidly it did not work. When this happened, he swapped the order of the elements, or left gaps. When gaps were left he made detailed predictions about what the new elements would be like.

When the elements were eventually discovered and their properties compared with the predictions Mendeleev had made, they matched very well and it proved how useful the periodic table was.

At the start of the twentieth century, scientists developed the technology required to discover protons, neutrons and electrons. When they looked back at Mendeleev's table, they realised that by occasionally swapping the order of the elements, what he had actually done was to place the elements in perfect order of increasing atomic number (number of protons).

In the modern periodic table, we say that the elements are arranged in order of increasing atomic number. Elements in the same group (column) have the same number of electrons in their outer shells.

**Group**

| | 1 | 2 | | | | | | | | | | 3 | 4 | 5 | 6 | 7 | 0 |
|---|---|---|---|---|---|---|---|---|---|---|---|---|---|---|---|---|---|
| **1** | | | | | | | H Hydrogen | | | | | | | | | | He Helium |
| **2** | Li Lithium | Be Beryllium | | | | | | | | | | B Boron | C Carbon | N Nitrogen | O Oxygen | F Fluorine | Ne Neon |
| **3** | Na Sodium | Mg Magnesium | | | | | | | | | | Al Aluminium | Si Silicon | P Phosphorous | S Sulfur | Cl Chlorine | Ar Argon |
| **4** | K Potassium | Ca Calcium | Sc Scandium | Ti Titanium | V Vanadium | Cr Chromium | Mn Manganese | Fe Iron | Co Cobalt | Ni Nickel | Cu Copper | Zn Zinc | Ga Gallium | Ge Germanium | As Arsenic | Se Selenium | Br Bromine | Kr Krypton |
| **5** | Rb Rubidium | Sr Strontium | Y Yttrium | Zr Zirconium | Nb Niobium | Mo Molybdenum | Tc Technetium | Ru Ruthenium | Rh Rhodium | Pd Palladium | Ag Silver | Cd Cadmium | In Indium | Sn Tin | Sb Antimony | Te Tellurium | I Iodine | Xe Xenon |
| **6** | Cs Caesium | Ba Barium | La Lanthanum | Hf Hafnium | Ta Tantalum | W Tungsten | Re Rhenium | Os Osmium | Ir Iridium | Pt Platinum | Au Gold | Hg Mercury | Tl Thallium | Pb Lead | Bi Bismuth | Po Polonium | At Astatine | Rn Radon |
| **7** | Fr Francium | Ra Radium | Ac Actinium | | | | | | | | | | | | | | | |

*Period*

# Mass Number and Atomic Number

Two numbers are often written next to the symbols from the periodic table. These are the mass number and the atomic number.

The **mass number** tells us the number of protons added to the number of neutrons in the nucleus of the atom.

The **atomic number**, which is sometimes called the proton number, tells us the number of protons in the nucleus of an atom. In a neutral atom this also tells us the number of electrons.

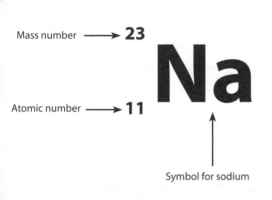

Mass number → **23**

**Na**

Atomic number → **11**

Symbol for sodium

For example, an atom of sodium has a mass number of 23 and an atomic number of 11. This means that this atom of sodium has:

● 11 protons (atomic number)
● 11 electrons (atomic number)
● 12 neutrons (mass number – atomic number).

An atom of oxygen has a mass number of 16 and an atomic number of 8. This means that this atom of oxygen has:

● 8 protons (atomic number)
● 8 electrons (atomic number)
● 8 neutrons (mass number – atomic number).

## Isotopes

Isotopes are different forms of the same element. This means that they have the same number of protons but a different number of neutrons.

## QUESTIONS

### QUICK TEST

1. What are the vertical columns in the periodic table called?

2. What are the horizontal rows in the periodic table called?

3. What does the mass number tell us?

4. What is the same about isotopes of an element?

5. What is different about isotopes of an element?

### EXAM PRACTICE

1. An isotope of carbon has a mass number of 13 and an atomic number of 6.

   How many protons, electrons and neutrons does this atom of carbon have?          **(1 mark)**

## SUMMARY

● **In the modern periodic table, the elements are placed in order of increasing atomic number.**

● **The atomic number is the number of protons in the nucleus of an atom.**

● **The mass number is the number of protons plus neutrons in the nucleus of an atom.**

● **Isotopes of an element have the same number of protons but a different number of neutrons.**

# Electronic Structure

## Atoms

Atoms consist of a small, central nucleus (that contains protons and neutrons) surrounded by shells of electrons.

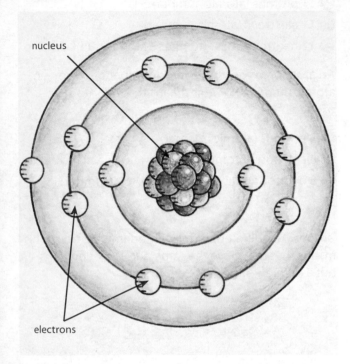

nucleus

electrons

## Electrons

Electrons fill up the shell closest to the nucleus (called the first shell) first. When this is full, they fill up the next shell.

In our model, there is room for up to two electrons in the first shell and up to eight electrons in the other shells.

**Example 1**

A lithium atom has three electrons – two in the first shell and one in the second shell.

This gives it an electronic structure of 2,1.

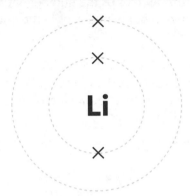

**Example 2**

A sodium atom has eleven electrons – two in the first shell, eight in the second shell and one in the third shell.

This gives it an electronic structure of 2,8,1.

The number of electrons in the outer shell of an atom reveals the group of the periodic table that the element belongs to.

Lithium and sodium both have just one electron in their outer shell, so both belong to Group 1 of the periodic table.

### Example 3

A carbon atom has six electrons – two in the first shell and four in the second shell.

This gives it an electronic structure of 2,4.

Carbon belongs to Group 4 of the periodic table.

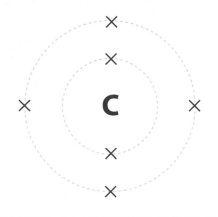

### Example 4

An oxygen atom has eight electrons – two in the first shell and six in the second shell.

This gives it an electronic structure of 2,6.

Oxygen belongs to Group 6 of the periodic table.

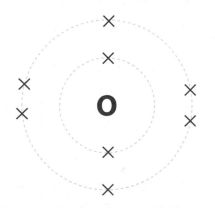

## SUMMARY

- Electrons are found in shells around the nucleus of an atom.

- In our model there are up to two electrons in the first shell and up to eight electrons in the other shells.

- The number of electrons in the outer shell tells us the group of the periodic table that the element belongs to.

## QUESTIONS

### QUICK TEST

1. Where are electrons found?

2. How many electrons can fit into the first shell?

3. What is the electronic structure of lithium?

4. What group does sodium belong to?

5. What group does carbon belong to?

### EXAM PRACTICE

1. An atom of magnesium has 12 electrons.

   a) What is the electronic structure of magnesium? **(1 mark)**

   b) What group does magnesium belong to? **(1 mark)**

# Ionic Bonding

## Ions

Ionic compounds are held together by strong forces of attraction between oppositely charged ions, called ionic bonds.

An **ion** is an atom or group of atoms with a charge.

Ions have a full, outer shell of electrons (like an atom of a noble gas).

Ionic bonding involves the transfer of **electrons** between atoms to form charged ions.

Atoms that gain electrons become negatively charged, while atoms that lose electrons become positively charged.

Ionic **bonding** is the attraction between these oppositely charged ions.

## Sodium Chloride

Sodium reacts with chlorine to form sodium chloride. The sodium atom transfers its outer electron to a chlorine atom. Both the sodium atom and the chlorine atom now have a full, outer shell of electrons.

The sodium atom that lost an electron now has a 1+ charge and is called a sodium ion. The chlorine atom has gained an electron and now has a 1− charge and is called a chloride ion.

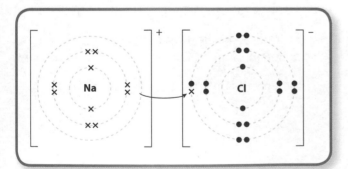

## Magnesium Oxide

Magnesium reacts with oxygen to form magnesium oxide. The magnesium atom transfers its two outer electrons to an oxygen atom.

The magnesium atom that lost two electrons now has a 2+ charge and is called a magnesium ion. The oxygen atom has gained two electrons and now has a 2− charge and is called an oxide ion.

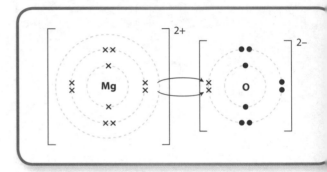

## Calcium Chloride

Calcium reacts with chlorine to form calcium chloride. The calcium atom transfers its two outer electrons to two chlorine atoms.

The calcium atom that lost two electrons now has a 2+ charge and is called a calcium ion. The chlorine atoms have gained one electron each and now have 1− charge and are called chloride ions.

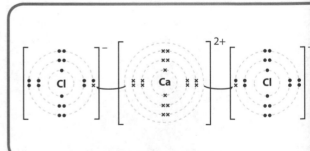

# Ionic Structures

Ionic compounds have **giant structures**. They are held together by strong forces of attraction between oppositely charged ions. These forces act in all directions. This means that ionic compounds have very high melting and boiling points because lots of energy must be supplied to break these strong forces of attraction.

Ionic compounds are solid at room temperature. Ionic solids do not conduct electricity because the ions are not able to move, but they do conduct when molten or when they are dissolved in something else.

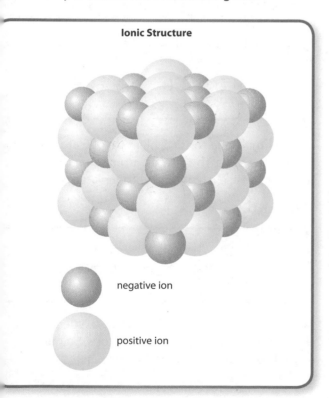

**Ionic Structure**

negative ion

positive ion

## Transition Metal Ions

When transition metals form compounds, the transition metal ions can have variable charges.

In copper(II) oxide, CuO, the copper ions have a 2+ charge.

In copper(I) oxide, $Cu_2O$, the copper ions have a 1+ charge.

## SUMMARY

● Ionic bonding is the attraction between oppositely charged ions.

● Sodium chloride, magnesium oxide and calcium chloride are all ionic compounds.

● Ionic compounds have high melting and boiling points. They conduct electricity when they are molten or dissolved in something else.

## QUESTIONS

### QUICK TEST

1. What is an ion?

2. What is the charge on an ion formed when an atom gains one electron?

3. What is the charge on an ion formed when an atom loses one electron?

4. What is the charge on an ion formed when an atom gains two electrons?

5. What is the charge on an ion formed when an atom loses two electrons?

### EXAM PRACTICE

1. Sodium reacts with chlorine to form an ionic compound.

   a) Why does this compound have a high melting point? **(2 marks)**

   b) Write a word equation for the reaction. **(1 mark)**

   c) Sodium atoms lose an electron to form sodium ions. What is the charge on a sodium ion? **(1 mark)**

# Covalent Bonding

## Covalent Bonds

In covalent bonding, atoms gain a full outer shell by sharing pairs of electrons.

A covalent bond is a shared pair of **electrons**. Non-metal atoms can gain a fuller, outer shell of electrons by **sharing** pairs of electrons.

In a covalent bond there is an electrostatic attraction between the nuclei of the atoms and the bonding electrons.

## Hydrogen, $H_2$

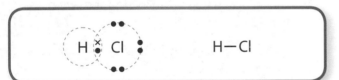

A hydrogen molecule is formed when two hydrogen atoms share a pair of electrons. Both atoms now have a full, outer shell of electrons.

## Hydrogen Chloride, HCl

A hydrogen chloride molecule is formed when a hydrogen atom and a chlorine atom share a pair of electrons. Both atoms now have a full, outer shell of electrons.

## Ammonia, $NH_3$

An ammonia molecule consists of a nitrogen atom surrounded by three hydrogen atoms. The nitrogen atom shares a pair of electrons with each of the hydrogen atoms.

## Methane, $CH_4$

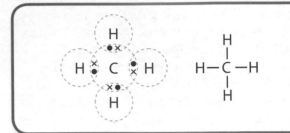

A methane molecule consists of a carbon atom surrounded by four hydrogen atoms. The carbon atom shares a pair of electrons with each of the hydrogen atoms.

## Water, $H_2O$

A water molecule consists of an oxygen atom and two hydrogen atoms. The oxygen atom shares a pair of electrons with each of the hydrogen atoms.

## Oxygen, $O_2$

An oxygen molecule consists of two oxygen atoms. The atoms share two pairs of electrons. So they are joined by a double covalent bond.

# Types of Structures

There are two types of covalent structure:

- Simple covalent structures
- Giant covalent structures.

## Simple Covalent Structures (Simple Molecules)

Simple covalent structures are formed by small numbers of atoms. There are very strong forces of attraction within these molecules, but much weaker forces of attraction between one molecule and another. This means that simple covalent structures have low melting and boiling points, and most are liquids or gases at room temperature. They do not conduct electricity because they do not contain ions or free electrons.

## Giant Covalent Structures (Macromolecular)

Giant covalent structures include:

- diamond
- graphite
- silicon dioxide.

These structures consist of very large numbers of atoms. All the atoms are held together by strong covalent bonds, so they have high melting and boiling points, and are solid at room temperature.

They do not conduct electricity (except graphite) because they do not contain ions and they do not have free electrons. They are insoluble in water.

## SUMMARY

- A covalent bond is a shared pair of electrons.
- Simple covalent structures contain a small number of atoms.
- Diamond, graphite and silicon dioxide are examples of giant covalent structures.

## QUESTIONS

### QUICK TEST

1. What is a covalent bond?
2. Describe the electrostatic attraction in a covalent bond.
3. Give the formula of a methane molecule.
4. Give the formula of an ammonia molecule.
5. Name a compound with a giant covalent structure.

### EXAM PRACTICE

1. Substance A has a boiling point of 80°C and does not conduct electricity. What type of structure does substance A have?

   Explain your answer. **(3 marks)**

# Alkali Metals

## Alkali Metals

The alkali metals belong to **Group 1** of the periodic table. They all have similar properties as they have one electron in their outer shell.

The alkali metals include:

- lithium
- sodium
- potassium.

They are found on the far left-hand side of the periodic table.

Group 1 metals have similar properties because they have similar electronic structures. Alkali metals react with non-metals to form ionic compounds. The alkali metal atom loses an electron to form an ion with a 1+ charge.

$$Na \rightarrow Na^+ + e^-$$

The alkali metal atom is oxidised.

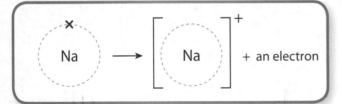

All the alkali metals are very **reactive**. They react vigorously with water to form a metal hydroxide and hydrogen. For this reason they are stored under oil.

**Example**

sodium + water → sodium hydroxide + hydrogen

$$2Na + 2H_2O \rightarrow 2NaOH + H_2$$

Lithium, sodium and potassium are all less dense than water. This means that if a small piece of metal is placed onto a trough of water, the metal floats on top of the water as it reacts.

LITHIUM

There is a gradual increase in reactivity down the group. This is because the outer electron is further from the nucleus, so it is lost more easily.

## Sodium and Chlorine

If sodium is burnt in chlorine gas, the compound sodium chloride is formed.

sodium + chlorine → sodium chloride

The reaction can be summed up as:

$$2Na(s) + Cl_2(g) \rightarrow 2NaCl(s)$$

- (s) indicates that the substance is a solid.
- (l) indicates that the substance is a liquid.
- (g) indicates that the substance is a gas.
- (aq) indicates that the substance is aqueous or dissolved in water.

Group 1 metals form white compounds that dissolve to form colourless solutions.

# Flame Tests

**Flame tests** can be used to identify the metals in compounds. The colour of the flame indicates the metal present:

- lithium – red
- sodium – orange
- potassium – lilac.

# Transition Metals

Transition metals are found in the middle of the periodic table. Compared with Group 1 metals, they are hard, strong and much less reactive.

With the exception of mercury, which is a liquid at room temperature, transition metals have higher melting points than Group 1 metals.

While Group 1 metal compounds are white, transition metals form coloured compounds. They are useful catalysts and are also used as pigments and dyes.

## SUMMARY

- Group 1 metals react vigorously with water.
- Sodium reacts with chlorine to make sodium chloride.
- Group 1 metals can be identified using flame tests.
- Group 1 metals are more reactive than transition metals.

## QUESTIONS

### QUICK TEST

1. Where are the alkali metals found in the periodic table?

2. Which of these metals would react most vigorously with water: sodium, lithium or potassium?

3. Name the gas produced when an alkali metal reacts with water.

4. What colour is seen when a flame test is carried out on a sodium compound?

5. What colour is seen when a flame test is carried out on a potassium compound?

### EXAM PRACTICE

1. Sodium reacts with water to form sodium hydroxide and hydrogen.

   a) Write a word equation for this reaction.
      **(1 mark)**

   b) Write a balanced symbol equation for this reaction. **(2 marks)**

   c) Why must sodium be stored under oil?
      **(1 mark)**

   d) When potassium reacts with water the reaction is more vigorous than when sodium reacts with water. Why does potassium react more vigorously? **(2 marks)**

# Noble Gases and Halogens

## Periods and Groups

The horizontal rows in the **periodic table** are called **periods**. The vertical columns are called **groups**. Elements in the same group have the same number of electrons in their outer shell, so they have similar chemical properties.

## The Noble Gases

The noble gases include:

- helium
- neon
- argon.

The noble gases are found on the far right-hand side of the periodic table. All the noble gases are very **unreactive** because they have a full, stable outer shell of electrons. This can make them very useful.

The following are some uses of the noble gases:

- Helium is **less dense** than air, so if a balloon is filled with helium gas it will float. Helium balloons are popular decorations at parties and special events.
- Neon is widely used in electrical discharge tubes.
- Argon is used to make filament lamps.

## The Halogens

The halogens include:

- fluorine
- chlorine
- bromine
- iodine.

The halogens are found next to the noble gases in the periodic table.

Chlorine is a pale green gas, bromine is a brown liquid and iodine is a dark grey solid. This shows us that the **boiling point** of the halogens increases as you go down the group.

Chlorine is used to sterilise water and in the manufacture of pesticides and plastics. Iodine is used to sterilise cuts.

The halogens react vigorously with alkali metals to form metal **halides**. The reaction between sodium and chlorine can be summed up by the following equation:

$$\text{sodium} + \text{chlorine} \rightarrow \text{sodium chloride}$$
$$2Na + Cl_2 \rightarrow 2NaCl$$

The halogen atom, chlorine, gains an electron to form a chloride ion, $Cl^-$. So the chlorine atom is reduced in this reaction.

Sodium chloride is used in food preparation as a flavouring (common salt) and as a preservative. It is also used in the production of chlorine gas.

here is a gradual **decrease** in **reactivity** as you go
down the group, so chlorine is more reactive than
bromine and iodine. This is because when an atom
reacts to form an ion, the electron is placed into a shell
increasingly further away from the nucleus.

A more reactive halogen will displace a less reactive
halogen from its solution. So chlorine will **displace**
bromine from a solution of potassium bromide.

| chlorine | + | potassium bromide | → | potassium chloride | + | bromine |
|---|---|---|---|---|---|---|
| $Cl_2$ | + | $2KBr$ | → | $2KCl$ | + | $Br_2$ |

## CFCs

In the past, chlorine was used in the manufacture of
CFCs. CFCs were chemically inert, had a low boiling
point and were insoluble in water. They were widely
used as refrigerants and aerosol propellants.

If CFCs escape into the atmosphere, they can break up
the ozone molecules that filter out the Sun's harmful
ultraviolet rays. Consequences of this include:

● a higher risk of sunburn
● skin aging faster
● increased risk of skin cancers and cataracts.

Today CFCs have been replaced by alkanes and HFCs
that will not damage the ozone layer.

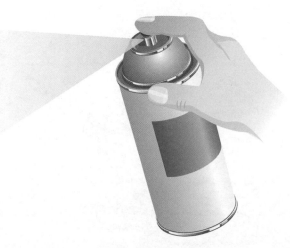

## SUMMARY

● **Noble gases are very unreactive and this makes them useful.**

● **Noble gas atoms have a full outer shell of electrons.**

● **Displacement reactions can be used to place the halogens into an order of reactivity.**

● **CFCs were used as aerosols and in fridges until it was discovered that they damaged the ozone layer.**

## QUESTIONS

### QUICK TEST

1. To which group does iodine belong?

2. To which group does neon belong?

3. In what state is bromine at room temperature?

4. What is helium used for?

5. What is argon used for?

### EXAM PRACTICE

1. The equation below shows the reaction
   between chlorine gas and a solution of
   potassium iodide.

   chlorine + potassium iodide → potassium chloride + iodine

   Explain how this reaction can be used to
   compare the reactivity of chlorine and iodine.
   **(2 marks)**

# Calculations 1

## Relative Atomic Mass

**Relative atomic mass** is used to compare the masses of different atoms. The relative atomic mass of an element is the weighted average mass of the isotopes of the element compared with an atom of carbon-12, which has a mass of 12.

## Relative Formula Mass

**Relative formula mass** is worked out by adding together the relative atomic masses of the atoms, in the ratio indicated by the chemical formula.

> **Example**
> What is the relative formula mass of water, $H_2O$?
>
> Relative atomic mass of hydrogen = 1
>
> Relative atomic mass of oxygen = 16
>
> The relative formula mass of water = $(1 \times 2) + (16 \times 1) = $ **18**

## Moles

The relative formula mass of a substance in grams is known as one mole of the substance, so one mole of water has a mass of 18 g.

Two moles of water have a mass of 36 g.

## Calculating the Mass of a Product

We can work out the theoretical yield of a reaction from a balanced symbol equation.

> **Example**
> Hydrogen reacts with oxygen to form water.
>
> $$2H_2 + O_2 \rightarrow 2H_2O$$
>
> If 4 g of hydrogen is burnt, what is the theoretical yield of water produced?
>
> The relative formula mass of $H_2 = 2$
>
> The relative formula mass of $H_2O = 18$
>
> The number of moles in 4 g of $H_2 = \dfrac{4\,g}{2\,g} = 2$ moles
>
> From the balanced symbol equation:
>
> 2 moles of hydrogen makes 2 moles of water.
>
> The mass of 2 moles of water = $2 \times 18 = 36$ g
>
> So 4 g of hydrogen would form **36 g of water**.

# Percentage Yield

The amount of product made in a reaction is called the **yield**. We often find that the actual yield of a reaction is lower than the theoretical yield.

This could be for a number of reasons:
- The reaction is reversible and does not go to completion.
- Some of the product is lost, for example, during filtering or evaporation.
- There may be side reactions that are producing another product.

$$\text{The percentage yield of a reaction} = \frac{\text{actual amount of product}}{\text{theoretical yield}} \times 100\%$$

### Example

The theoretical yield for a reaction is calculated as 1.2 g.

A student carries out the reaction but only produces 1.0 g of product.

The percentage yield $= \frac{1.0}{1.2} \times 100\% = \textbf{83.3\%}$

## Atom Economy

$$\text{The atom economy of a reaction} = \frac{\text{mass of useful product}}{\text{total mass of product}} \times 100\%$$

Scientists try to choose reactions that have a high atom economy.

### Example

In an experiment, 7.5 g of useful product was made. The total mass of the product was 10.0 g. What is the atom economy?

Atom economy $= \frac{7.5\,\text{g}}{10.0\,\text{g}} \times 100\% = \textbf{75\%}$

## SUMMARY

- The theoretical yield of a reaction can be worked out using relative atomic masses and relative formula masses.

- Percentage yield is a way of measuring the actual amount of product made in a reaction.

- Atom economy is a way of measuring the amount of useful product made in a reaction.

## QUESTIONS

### QUICK TEST

1. What is relative atomic mass used for?

2. What is the standard against which we measure the atomic mass of different atoms?

3. What is one mole equal to?

4. Why might a reversible reaction have a low percentage yield?

5. What is the equation for the percentage yield of a reaction?

### EXAM PRACTICE

1. A student calculates that the theoretical yield for her experiment is 2.5 g. She carries out the experiment and only produces 2.2 g of product.

   a) Why might her actual yield be less than her theoretical yield? **(1 mark)**

   b) What is the percentage yield of this reaction? **(1 mark)**

# Calculations 2

## Moles

Moles are used to measure the amount of a substance.

Moles can be calculated using the following formula:

$$\text{The number of moles of a substance} = \frac{\text{mass of the substance}}{\text{relative formula mass of the substance}}$$

This equation can be rearranged to give:

$$\text{Relative formula mass of the substance} = \frac{\text{mass of the substance}}{\text{number of moles}}$$

$$\text{Mass of the substance} = \text{number of moles} \times \text{relative formula mass of the substance}$$

### Example

Find the number of moles in 5.85 g of sodium chloride.

The relative formula mass of sodium chloride = 58.5

The number of moles $= \dfrac{5.85\,g}{58.5\,g} = 0.1$ moles

The number of moles present in solution can also be calculated.

$$\text{Number of moles} = \frac{\text{volume (cm}^3)}{1000} \times \text{concentration}$$

This equation can be rearranged to give:

$$\text{Volume} = \frac{\text{number of moles} \times 1000}{\text{concentration}}$$

$$\text{Concentration} = \frac{\text{number of moles} \times 1000}{\text{volume}}$$

## Volumetric Calculations

We can carry out a **titration** to find the concentration of a solution.

In a titration, a known volume of an alkaline solution is placed into a flask. An **indicator** is added. We can use indicators to find out exactly how much acid is required.

The indicator changes colour when exactly the right amount of acid has been added. The type of indicator used depends on the type of acid and the type of alkali involved.

If you are using a strong acid and a strong alkali, then any indicator would work well.

If you are using a strong acid and a weak alkali, then **methyl orange** should be used.

If you are using a weak acid and a strong alkali, then **phenolphthalein** should be used.

No indicators work well for a weak acid and a weak alkali.

- Strong acids include hydrochloric acid, nitric acid and sulfuric acid.
- Strong bases include sodium hydroxide and potassium hydroxide.
- Weak acids include carboxylic acids like ethanoic acids.
- Weak bases include ammonium hydroxide or 'ammonia'.

## Example

$$HCl + NaOH \rightarrow NaCl + H_2O$$

20.5 cm³ of hydrochloric acid, which has a concentration of 0.1 mol dm⁻³, was required to neutralise 25 cm³ of sodium hydroxide solution.

What is the concentration of the sodium hydroxide solution?

The number of moles in 20.5 cm³ of 0.1 mol dm⁻³ hydrochloric acid

$$= \frac{20.5}{1000} \times 0.1 = 0.00205$$

From the equation, 0.00205 moles of hydrochloric acid will react with 0.00205 moles of sodium hydroxide.

The number of moles in 25 cm³ of alkali

$$= 0.00205$$

The concentration of the alkali

$$= \frac{\text{number of moles} \times 1000}{\text{volume}}$$

$$= \frac{0.00205 \times 1000}{25}$$

$$= 0.082 \, \text{mol dm}^{-3}$$

The sodium hydroxide solution has a concentration of **0.082 mol dm⁻³**.

## SUMMARY

- Moles are used to measure the amount of a substance.

- Indicators can be used to identify acidic, alkaline and neutral solutions.

- There are strong and weak acids and alkalis.

## QUESTIONS

### QUICK TEST

1. Carbon dioxide has a formula mass of 44.

   a) How many moles are present in 2.2 g of carbon dioxide?

   b) How many moles are present in 8.8 g of carbon dioxide?

   c) How many moles are present in 11 g of carbon dioxide?

2. A solution of hydrochloric acid has a concentration of 0.1 mol dm⁻³.

   a) How many moles are present in 25 cm³ of this solution?

   b) How many moles are present in 100 cm³ of this solution?

### EXAM PRACTICE

1. Name the indicator you would use to titrate the weak acid ethanoic acid with the strong alkali sodium hydroxide. **(1 mark)**

# Calculations 3

## Empirical Formula

The empirical formula of a compound is the simplest whole number ratio of atoms in the formula.

Ethene has the molecular formula $C_2H_4$ so it has an empirical formula of $CH_2$.

We can also use reacting masses to calculate the empirical formula of a compound.

### Example

Magnesium reacts with oxygen to form the compound magnesium oxide.

6 g of magnesium reacts with 4 g of oxygen. What is the empirical formula of magnesium oxide?

|  | Mg | O |
|---|---|---|
| **Mass** | 6 | 4 |
| Find the number of moles: $\frac{\text{mass}}{\text{relative atomic mass}}$ | $\frac{6}{24}$ | $\frac{4}{16}$ |
| This is the ratio in which the atoms combine. | 0.25 | 0.25 |
| Divide through by the smallest number… | $\frac{0.25}{0.25}$ | $\frac{0.25}{0.25}$ |
| … to get the ratio of atoms in its simplest terms. | 1 | 1 |

The empirical formula of magnesium oxide is **MgO**.

## Bond Enthalpy Calculations

Energy is released when **bonds** are made and energy is taken in when bonds are broken.

Overall, **exothermic** reactions give out energy and **endothermic** reactions take in energy.

**Bond energy** is the average amount of energy that must be taken in to break one mole of that bond. We can use bond energies to work out whether a reaction is exothermic or endothermic.

### Example

Hydrogen reacts with chlorine to form hydrogen chloride. Is this reaction exothermic or endothermic?

hydrogen + chlorine → hydrogen chloride

$$H_2 \quad + \quad Cl_2 \quad \rightarrow \quad 2HCl$$

|  | Bond energy (kJ mol$^{-1}$) |
|---|---|
| H–H | 436 |
| Cl–Cl | 242 |
| H–Cl | 431 |

The energy taken in to break the old bonds:

1 mole of H–H = 436 kJ

1 mole of Cl–Cl = 242 kJ

Total = 678 kJ

The energy released when the new bonds are made:

2 moles of H–Cl = 2 × 431 = 862 kJ

Overall, more energy is released when the new bonds are formed than is taken in to break the old bonds so the reaction is **exothermic**.

## Avogadro's Number

One mole of any substance contains approximately
$\times 10^{23}$ particles. The term 'particles' could refer to
atoms, molecules or ions.

## Molar Volume

One mole of any gas at a given temperature takes up
the same volume.

At a temperature of 25 °C and a pressure of
atmosphere, one mole of any gas will occupy
$4 \, dm^3$.

This means that one mole of nitrogen would occupy a
volume of 24 dm³. Two moles of oxygen would occupy
volume of 48 dm³. Half a mole of hydrogen would
ke up a volume of 12 dm³.

## SUMMARY

- The empirical formula is the simplest whole number ratio of atoms in the formula of a substance.

- Energy is released when bonds are made and taken in when bonds are broken.

- Avogadro's number is the number of particles in one mole of any substance.

- One mole of any gas occupies a volume of 24 dm³ at room temperature and pressure.

## QUESTIONS

### QUICK TEST

1. What is the empirical formula of a compound?

2. What is the empirical formula of $C_3H_6$?

3. What is the empirical formula of $N_2H_4$?

4. What are the units used to measure bond energy?

5. During a reaction, overall energy is given out. What type of reaction is this?

### EXAM PRACTICE

1. methane + oxygen → carbon dioxide + water

$$CH_4 + 2O_2 \rightarrow CO_2 + 2H_2O$$

When one mole of methane is burnt in oxygen, 2644 kJ is required to break the bonds in methane and oxygen.

3338 kJ of energy is released when the bonds in carbon dioxide and water are made.

Is this reaction exothermic or endothermic? Explain your answer.

**(2 marks)**

# The Haber Process

## Making Ammonia

The Haber process is used to make ammonia, $NH_3$. Nitrogen (from the fractional distillation of liquid air) is reacted with hydrogen (from natural gas) in a reversible reaction.

The reaction can be summed up as:

> nitrogen + hydrogen ⇌ ammonia
>
> $N_2(g)$ + $3H_2(g)$ ⇌ $2NH_3(g)$

The state symbol (g) means gaseous – a gas.

The forward reaction is exothermic.

On cooling, the ammonia liquefies and can be removed. Any unreacted nitrogen and hydrogen are recycled. The ammonia is produced all the time in a continuous process. Ammonia is used in the manufacture of fertilisers and cleaning fluids.

In a closed system, eventually an equilibrium is reached. The relative amount of the substances at equilibrium depends on the conditions.

The conditions chosen in the Haber process are typically

- An iron catalyst.
- A high pressure of around 200 atmospheres.
- A moderate temperature of around 450 °C.

An iron catalyst increases the rate of reaction. Catalys are often used in industry because they allow us to use less energy, which makes the process cheaper.

A high pressure is used to increase the yield of ammonia. Increasing the pressure favours the forwar reaction, which has fewer gas molecules on the product side.

The forward reaction is exothermic, so a high temperature would give a good rate of reaction but a poor yield of ammonia. A low temperature would giv a poor rate of reaction but a good yield of ammonia. practice, a compromise temperature is used that give a reasonable rate and a reasonable yield.

**The Haber Process**

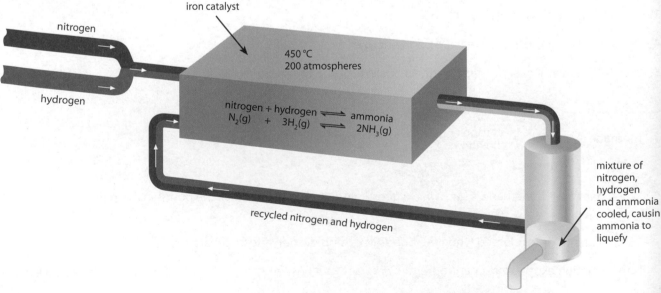

iron catalyst

nitrogen

450 °C
200 atmospheres

hydrogen

nitrogen + hydrogen ⇌ ammonia
$N_2(g)$ + $3H_2(g)$ ⇌ $2NH_3(g)$

recycled nitrogen and hydrogen

mixture of nitrogen, hydrogen and ammonia cooled, causin ammonia to liquefy

liquid ammonia

# Fertilisers

Fertilisers help crops to grow bigger and faster. They replace the essential elements (including nitrogen, phosphorus and potassium) used by plants as they grow. Plants absorb these elements through their roots.

Many fertilisers can be made by a neutralisation reaction between an acid and an alkali.

Ammonia can be oxidised and then reacted with water to form nitric acid. This can be reacted with ammonia to form the popular fertiliser ammonium nitrate.

$$\text{nitric acid} + \text{ammonia} \rightarrow \text{ammonium nitrate}$$
$$HNO_3 + NH_3 \rightarrow NH_4NO_3$$

In a similar way, reacting ammonia with sulfuric acid forms ammonium sulfate, while reacting ammonia with phosphoric acid forms ammonium phosphate.

## Percentage by Mass

Plants use nitrogen to produce protein. Being able to calculate the percentage of nitrogen in a fertiliser allows us to choose the right amount of fertiliser to use.

Percentage mass of an element in a compound =

$$\frac{\text{relative atomic mass of the element} \times \text{number of atoms}}{\text{relative formula mass of the compound}} \times 100\%$$

**Example**
Work out the percentage of nitrogen in ammonium nitrate, $NH_4NO_3$.

$$= \frac{14 \times 2}{80} \times 100\%$$

$$= 35\%$$

So the percentage of nitrogen in ammonium nitrate is **35%**.

## SUMMARY

- Ammonia is made from nitrogen and hydrogen.

- An iron catalyst, a pressure of 200 atmospheres and a temperature of 450°C are used in the Haber process.

- Ammonia is used to make fertilisers.

- The percentage mass of an element in a compound tells us about the composition of the compound.

## QUESTIONS

### QUICK TEST

1. Where is nitrogen obtained from?

2. Where is hydrogen obtained from?

3. What is the catalyst used in the Haber process?

4. What is the temperature used in the Haber process?

5. What is the pressure used in the Haber process?

### EXAM PRACTICE

1. Ammonia is a very useful chemical. It is made using the Haber process, which involves a reversible reaction between nitrogen and hydrogen. Give the symbol equation for this reaction. Describe the conditions under which ammonia is made and explain why these conditions are used.

   *The quality of your written communication will be assessed in your answer to this question.* **(6 marks)**

# The Manufacture of Sulfuric Acid

## Dynamic Equilibrium

Some reactions are reversible. If a reversible reaction is carried out in a closed system (where nothing can enter or leave), then eventually a dynamic equilibrium will be reached.

In a **dynamic equilibrium** both the forwards and the reverse reactions are still happening, but they happen at the same rate.

Conditions can affect the position of equilibrium. In industrial processes we control the conditions to make the maximum profit.

## The Contact Process

The contact process is used to produce sulfuric acid, $H_2SO_4$.

### Stage 1

Sulfur is burnt in oxygen to form sulfur dioxide.

$$\text{sulfur} \quad + \quad \text{oxygen} \quad \rightarrow \quad \text{sulfur dioxide}$$
$$\text{S} \quad + \quad O_2 \quad \rightarrow \quad SO_2$$

### Stage 2

The sulfur dioxide is reacted with more oxygen over a **vanadium(v) oxide** catalyst at a temperature of around 450 °C and a pressure of 2 atmospheres.

$$\text{sulfur dioxide} \quad + \quad \text{oxygen} \quad \rightleftharpoons \quad \text{sulfur trioxide}$$
$$2SO_2 \quad + \quad O_2 \quad \rightleftharpoons \quad 2SO_3$$

This reaction is reversible and the forward reaction is **exothermic**.

The vanadium(v) oxide catalyst increases the rate of reaction but does not affect the position of equilibrium.

A moderate temperature of around 450 °C is used. A higher temperature would give a better rate of reaction but a poorer yield of sulfur trioxide. A lower temperature would give a better yield of sulfur trioxide but a poorer rate of reaction. A moderate temperature gives a reasonable rate of reaction and a reasonable yield of sulfur trioxide.

A pressure of 2 atmospheres is used. A higher pressure would give a better yield of sulfur trioxide but the yield is already so high that it is not economically worthwhile to use a very high pressure.

A pressure of 2 atmospheres is enough to keep the gases moving through the system.

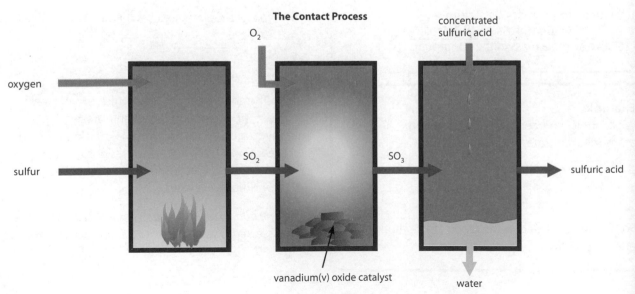

**The Contact Process**

oxygen · sulfur · $O_2$ · $SO_2$ · $SO_3$ · vanadium(v) oxide catalyst · water · concentrated sulfuric acid · sulfuric acid

## Stage 3

The sulfur trioxide cannot be reacted directly with water because the reaction would produce a hazardous fog of sulfuric acid droplets.

Instead the sulfur trioxide is reacted with concentrated sulfuric acid, which has a concentration of 98%.

The sulfur trioxide reacts with the water molecules in the acid to form an even more concentrated sulfuric acid. The acid now has a concentration of about 99.5%. This acid is then diluted.

Some acid is retained to produce more sulfuric acid in the future and some is sold to customers.

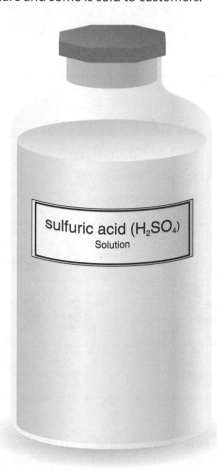

sulfuric acid (H₂SO₄)
Solution

## SUMMARY

- In a dynamic equilibrium the rate of forward reaction is equal to the rate of backward reaction.

- Sulfuric acid is made using the contact process.

## QUESTIONS

### QUICK TEST

1. What is a closed system?

2. What is a dynamic equilibrium?

3. What catalyst is used in the manufacture of sulfuric acid?

4. What temperature is used in the manufacture of sulfuric acid?

5. What pressure is used in the manufacture of sulfuric acid?

### EXAM PRACTICE

1. In the production of sulfuric acid, sulfur dioxide is reacted with oxygen to form sulfur trioxide.

   sulfur dioxide + oxygen $\rightleftharpoons$ sulfur trioxide

   $$2SO_2 \quad + \quad O_2 \quad \rightleftharpoons \quad 2SO_3$$

   a) How is sulfur dioxide produced?  **(1 mark)**

   b) What does the symbol $\rightleftharpoons$ mean?  **(1 mark)**

   c) A catalyst called vanadium(v) oxide is used in this reaction. How does this catalyst affect the rate of reaction and the position of equilibrium?  **(2 marks)**

# Rates of Reaction

## Increasing the Rate of Reaction

### Temperature

The rate of reaction tells us how fast a chemical reaction takes place. A chemical reaction takes place when reacting particles **collide** and have enough **energy** to react – this is called **activation energy**.

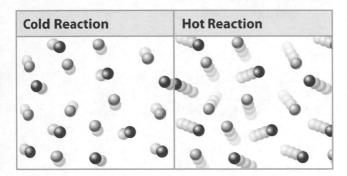

When the **temperature** is increased:

1. Particles move faster.
2. The particles collide more often.
3. When the particles collide they have more energy, so more particles have enough energy to react.
4. The rate of reaction increases.

### Concentration

When the **concentration** of a solution is increased:

1. The particles are closer together.
2. The particles collide more often
3. The rate of reaction increases.

Concentration is measured in moles per cubic decimetre, mol dm$^{-3}$.

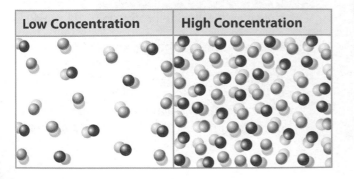

### Pressure

When the **pressure** of gases is increased:

1. The particles are closer together.
2. The particles collide more often.
3. The rate of reaction increases.

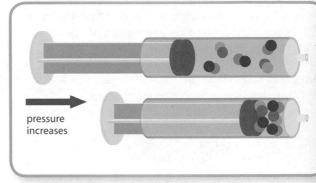

pressure increases

### Surface Area

When the **surface area** of a solid is increased (small pieces have a large surface area):

1. The particles collide more often.
2. The rate of reaction increases.

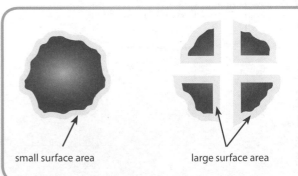

small surface area          large surface area

### Catalyst

Adding a **catalyst** increases the rate of a chemical reaction. Catalysts are specific to certain reactions. They are not used up during reactions so they can be reused many times. Catalysts are used in industry to reduce production costs. Enzymes are biological catalysts.

# Following a Reaction

When magnesium metal is reacted with dilute hydrochloric acid, a salt called magnesium chloride and the gas hydrogen are made.

$$magnesium + \frac{hydrochloric}{acid} \rightarrow \frac{magnesium}{chloride} + hydrogen$$

We can follow how fast this reaction happens by measuring:

- how quickly the hydrogen gas is made
- how quickly the mass of the reaction flask goes down as the hydrogen gas is made and escapes from the flask.

This graph shows the amount of hydrogen produced in two experiments.

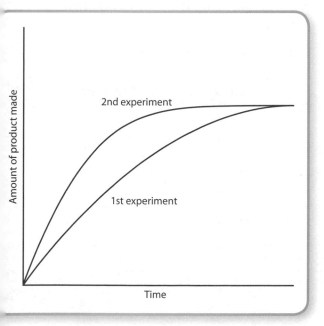

In both experiments the rate of reaction is fastest at the start of the reaction because the concentration of the reactants is highest at the start.

The reaction is over when the line levels out because one of the reactants has been used up.

The second experiment has a faster rate of reaction than the first experiment.

## QUESTIONS

### QUICK TEST

1. What needs to happen for two particles to react?

2. How does increasing temperature affect the rate of a chemical reaction?

3. How does decreasing the concentration of a solution affect the rate of a chemical reaction?

4. How does adding a catalyst affect the rate of a chemical reaction?

5. During an experiment, when is the rate of a chemical reaction fastest?

### EXAM PRACTICE

1. When calcium carbonate reacts with hydrochloric acid, the following are produced: calcium chloride, water and carbon dioxide. The rate of reaction can be found by measuring how much carbon dioxide is produced every 10 seconds. Suggest how and why the rate of this reaction would change if:
   - The temperature was increased.
   - The concentration of the hydrochloric acid was increased.

   *The quality of your written communication will be assessed in your answer to this question.*

   **(6 marks)**

# Energy

## Exothermic Reactions

In **exothermic** reactions, energy (normally in the form of heat) is transferred to the surroundings. This means that if we recorded the temperature change during the reaction, we would see a temperature increase.

The burning of fuels, rusting and neutralisation are all examples of exothermic reactions. Most reactions are exothermic.

The fuel used in Bunsen burners is called methane. When methane is burnt it reacts with oxygen to release heat energy.

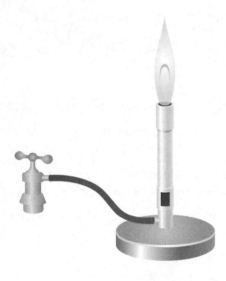

Neutralisation reactions are also exothermic. If sodium hydroxide solution is reacted with hydrochloric acid, heat energy is released.

## Endothermic Reactions

In **endothermic** reactions, energy (normally in the form of heat) is taken in from the surroundings. This means that if we recorded the temperature change during the reaction, we would see a temperature decrease.

Thermal decomposition reactions and photosynthesis are endothermic.

## Bonds

Energy must be supplied to break **bonds**.

Energy is released when new bonds are made.

In exothermic reactions, more energy is released when new bonds are formed than is taken in to break old bonds.

In endothermic reactions, more energy is taken in to break the old bonds than is released when the new bonds are formed.

The table below shows some average bond energies.

| Bond | Bond Energy kJ mol$^{-1}$ |
|------|---------------------------|
| C–C | 347 |
| O=O | 498 |
| C–H | 413 |
| O–H | 464 |
| C–O | 358 |
| H–Cl | 432 |

# Reversible Reactions

Some reactions are **reversible**. They can go in a forward or in a reverse direction. If A and B are reactants and C and D are products, then a reversible reaction can be summed up by:

$$A + B \rightleftharpoons C + D$$

If the forward reaction is exothermic, the reverse reaction is endothermic and the same amount of energy is transferred in each case.

When blue, hydrated copper(II) sulfate is heated, it decomposes to form white, anhydrous copper(II) sulfate and water. This reaction is endothermic. If water is added to anhydrous copper(II) sulfate, it forms hydrated copper sulfate. This reaction is exothermic and can be used to test that a liquid is really water.

These reactions can be summed up by the following equation:

$$\text{hydrated copper sulfate} \rightleftharpoons \text{anhydrous copper sulfate} + \text{water}$$

Ammonia is an important chemical. It is used to make fertilisers. Hydrogen reacts with nitrogen to form ammonia. The reaction is reversible and can be summed up by the following reaction:

$$\text{nitrogen} + \text{hydrogen} \rightleftharpoons \text{ammonia}$$

The forward reaction is exothermic, while the reverse reaction is endothermic.

## SUMMARY

- In exothermic reactions, energy is given out.
- In endothermic reactions, energy is taken in.
- New bonds are made and old bonds are broken in chemical reactions.
- Reversible reactions can go forwards or backwards.

## QUESTIONS

### QUICK TEST

1. Is the burning of a fuel an exothermic or endothermic reaction?

2. If there is a temperature decrease during a chemical reaction, what type of reaction has taken place?

3. What is released when a new bond is made?

4. What does the symbol $\rightleftharpoons$ indicate?

5. What type of reaction is thermal decomposition?

### EXAM PRACTICE

1. When hydrated copper(II) sulfate is heated, it forms anhydrous copper sulfate and water. The reaction is reversible.

   a) Write a word equation for this reaction. **(2 marks)**

   b) What would you **see** when hydrated copper sulfate is heated? **(1 mark)**

   c) How could you prove that a colourless liquid was really water? **(1 mark)**

# Calorimetry

## Measuring Energy Changes

We can measure the energy released when fuels are burned using a technique called **calorimetry**.

In this process the following takes place:

1.  The volume of the water in the boiling tube is measured.
2.  The temperature of the water in the boiling tube is measured.
3.  The mass of the spirit burner and fuel is recorded.
4.  The spirit burner is lit and the heat energy that is released as the fuel burns warms up the water in the boiling tube.
5.  The spirit burner is turned off when 1 g of fuel has been burned.
6.  The new temperature of the water is measured.

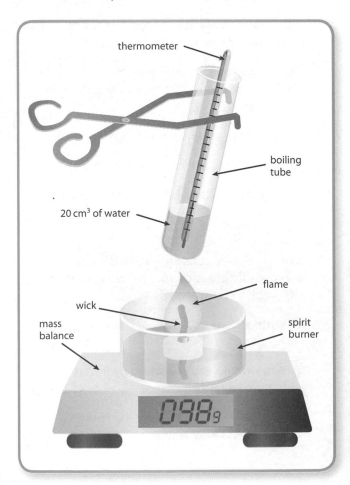

thermometer

boiling tube

20 cm³ of water

flame

wick

spirit burner

mass balance

098ɡ

## Calculating the Energy Transferred

We can calculate the heat energy transferred using the following equation:

$$\begin{array}{c}\text{heat} \\ \text{energy} \\ \text{transferred} \\ \text{(J)}\end{array} = \begin{array}{c}\text{mass of} \\ \text{water} \\ \text{(g)}\end{array} \times \begin{array}{c}\text{specific heat} \\ \text{capacity} \\ \text{of water} \\ \text{(J g}^{-1}\text{°C}^{-1}\text{)}\end{array} \times \begin{array}{c}\text{change in} \\ \text{temperature} \\ \text{(°C)}\end{array}$$

The specific heat capacity of water is $4.2 \, \text{J g}^{-1}\text{°C}^{-1}$

This means that it takes 4.2 J of energy to raise the temperature of 1 g of water by 1 °C.

> **Example**
>
> In a calorimetry experiment, 1 g of fuel raised the temperature of 10 g of water by 5 °C. What is the heat energy transferred?
>
> The heat energy transferred (J) = 10 g × $4.2 \, \text{J g}^{-1}\text{°C}^{-1}$ × 5 °C
>
> **= 210 J**

To compare the amount of energy released by different fuels, we can divide the heat energy transferred by the mass of fuel burned.

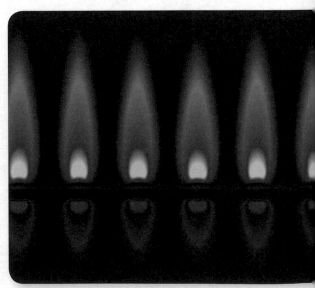

# Energy in Fuels

The table below shows the amount of energy released when 1 g of three fuels were burned.

| Fuel | Energy released (kJ) |
|------|----------------------|
| Hydrogen | 143 |
| Petrol | 48 |
| Ethanol | 30 |

Hydrogen is a gaseous fuel. It releases most energy per gram, but because it is a gas, large volumes of hydrogen are required.

Petrol is a liquid but is a non-renewable fuel. Ethanol is also a liquid, but it releases less energy per gram than petrol. It is a renewable fuel.

# Energy in Foods

Food labels display nutritional information. They tell us the amount of energy that the food contains. This is important because it helps us to make sure we get the right amount of energy from the food we eat. If we consume either too much energy or not enough energy, we could become ill.

In science we normally measure the amount of energy in joules. For historical reasons we normally measure the energy in foods in calories. One calorie is equal to 4.2 joules.

TOMATO SOUP
NUTRITION INFORMATION
Typical Values

| | Per 100g |
|---|---|
| Energy | 182kJ |
| Protein | 0.8g |
| Carbohydrates | 6.0g |
| Fat | 1.8g |
| Fibre | 0.6g |
| Salt | 0.3g |

INGREDIENTS
Tomatoes (81%), Water, Vegetable Oil, Sugar, Modified Cornflour, Salt, Dried Skimmed Milk, Cream, Spice Extracts, E120 (Colour), E200 (Preservative) Herbs, E330 Citric Acid (Antioxidant).

## SUMMARY

- Calorimetry experiments are used to measure the energy change in chemical reactions.
- Scientists compare the energy released when fuels are burnt.
- Food labels show the amount of energy in a serving of the food.

## QUESTIONS

### QUICK TEST

1. What is the name of the technique used to measure the amount of energy released when fuels are burned?

2. What is the unit used to measure the amount of heat energy transferred?

3. What is the unit used to measure the change in temperature?

4. What is the unit used to measure the mass of water?

5. What is one calorie equal to?

### EXAM PRACTICE

1. In a calorimetry experiment, 1 g of fuel raised the temperature of 5 g of water by 4 °C.

   a) Is the burning of this fuel an exothermic or an endothermic reaction? **(1 mark)**

   b) Calculate the heat energy transferred during this reaction. **(2 marks)**

# Energy Profile Diagrams

## Exothermic Reactions

We use energy profile diagrams to show the energy changes that happen in exothermic and endothermic reactions.

In exothermic reactions, the products contain less energy than the reactants, so energy is released by the reaction.

Activation energy is the amount of energy required to break the bonds in the reactants and get the reaction started.

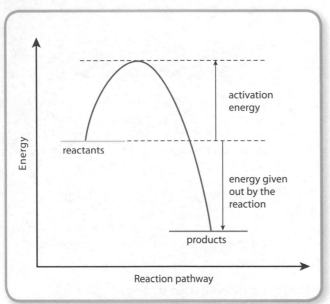

The burning of fuels, neutralisation and rusting are all exothermic reactions.

## Endothermic Reactions

In endothermic reactions, the products contain more energy than the reactants, so energy is taken in from the surroundings.

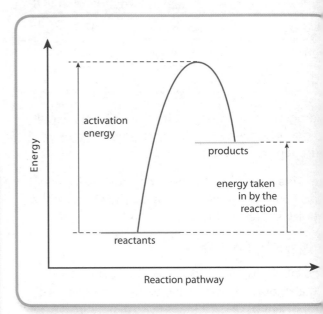

The thermal decomposition of calcium carbonate is a endothermic reaction.

# Catalysts

Catalysts are substances that speed up the rate of a chemical reaction but are not used up themselves. Catalysts work by offering an alternative reaction pathway with lower activation energy.

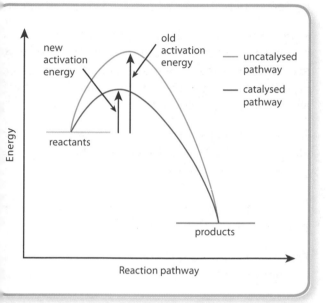

## SUMMARY

- Energy profile diagrams show the energy changes that take place in chemical reactions.

- Activation energy is the amount of energy needed to get the reaction started.

- Catalysts offer an alternative reaction pathway with lower activation energy.

## QUESTIONS

### QUICK TEST

1. What is taken in when bonds are broken?

2. What is activation energy?

3. What is given out during an exothermic reaction?

4. What do catalysts do?

5. Why can catalysts be reused?

### EXAM PRACTICE

1. Iron is used as a catalyst in the Haber process, which is used to make ammonia.

$$N_2 + 3H_2 \rightleftharpoons 2NH_3$$

The forward reaction is exothermic.

a) Draw an energy profile diagram for the forward reaction. **(4 marks)**

b) Describe, in terms of activation energy, how iron increases the rate of the reaction. **(1 mark)**

# Aluminium and Dilute Sulfuric Acid

## Aluminium

Pure aluminium has a low **density** but is too soft for many uses.

Aluminium can be mixed with other metals to form **alloys**. These alloys combine low density with high strength, and can be used, for example, in bike frames.

## Extraction of Aluminium

Aluminium is quite a reactive metal. It is extracted from its ore, **bauxite**, by **electrolysis**. This is an expensive process because it involves lots of steps and requires lots of energy.

Bauxite contains aluminium oxide. For electrolysis to occur, the aluminium ions and the oxide ions must be able to move. So, solid aluminium oxide must either be heated until it melts or dissolved in something else.

Aluminium oxide has a very high melting point, so heating aluminium oxide until it melts would be very expensive. Fortunately, another aluminium ore called **cryolite** has a much lower melting point. In practice, aluminium oxide is dissolved in molten cryolite.

During electrolysis, the $Al^{3+}$ ions move to the negative electrode (cathode) where aluminium forms.

$$Al^{3+} + 3e^- \rightarrow Al$$

The aluminium ions are **reduced**.

The oxide ions, $O^{2-}$, move to the positive electrode (anode) where they react to form oxygen molecules.

$$2O^{2-} \rightarrow O_2 + 4e^-$$

The oxide ions are **oxidised**.

The oxygen reacts with the **graphite** anode to produce carbon dioxide, so these electrodes must be regularly replaced.

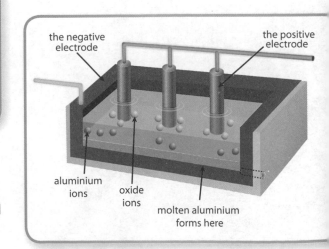

the negative electrode

the positive electrode

aluminium ions

oxide ions

molten aluminium forms here

# The Electrolysis of Dilute Sulfuric Acid

The electrolysis of dilute sulfuric acid produces **oxygen** and **hydrogen**.

Hydrogen is formed at the cathode.

$$2H^+ + 2e^- \rightarrow H_2$$

Oxygen is formed at the anode.

$$4OH^- \rightarrow 2H_2O + O_2 + 4e^-$$

## Gas Tests

| Hydrogen | Oxygen |
|---|---|
| Burns with a 'squeaky' pop | Relights a glowing splint |

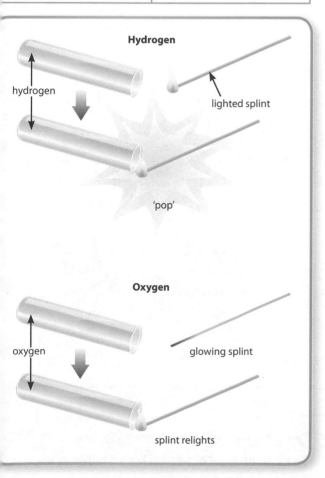

**Hydrogen**

hydrogen

lighted splint

'pop'

**Oxygen**

oxygen

glowing splint

splint relights

## SUMMARY

- Aluminium is extracted from its ore, bauxite, by electrolysis.
- Pure aluminium is quite soft.
- The electrolysis of dilute sulfuric acid produces oxygen and hydrogen.

## QUESTIONS

### QUICK TEST

1. What is the chemical compound found in bauxite?

2. Bauxite is one ore of aluminium. Name another ore of aluminium.

3. Does solid aluminium oxide conduct electricity?

4. Does molten aluminium oxide conduct electricity?

5. In the electrolysis of aluminium oxide, what is formed at the negative electrode?

### EXAM PRACTICE

1. Aluminium is extracted from aluminium oxide by electrolysis.

   a) During electrolysis, aluminium is deposited at the negative electrode. Give the symbol equation for this reaction. **(1 mark)**

   b) Oxygen is produced at the positive electrode. Give the symbol equation for this reaction. **(1 mark)**

   c) Why must the graphite electrodes be regularly replaced? **(1 mark)**

# Sodium Chloride

## Uses

Sodium chloride is a **salt**. Salts are used:

- in the production of fertilisers
- as colouring agents
- in fireworks.

## Sodium Chloride Solution

Sodium chloride (common salt) is a very important resource. It is found in large quantities dissolved in seawater and in underground deposits formed when ancient seas evaporated.

Rock salt (unpurified salt) is used during winter to grit roads, to stop them from becoming icy and dangerous.

**Rock Salt**

The **electrolysis** of concentrated sodium chloride solution produces:

- chlorine
- hydrogen
- sodium hydroxide.

During electrolysis, positive hydrogen, $H^+$, ions move to the negative electrode (cathode) where hydrogen is produced.

$$2H^+ + 2e^- \rightarrow H_2$$

Chloride ions, $Cl^-$, move to the positive electrode (anode) where chlorine is produced.

$$2Cl^- \rightarrow Cl_2 + 2e^-$$

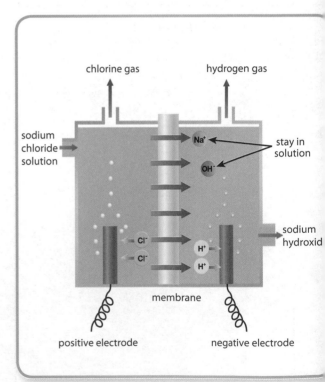

**Gas tests:**

- Hydrogen burns with a 'squeaky' pop.
- Chlorine bleaches damp litmus paper.

## ...ses of Sodium Chloride Solution

...he products of the electrolysis of concentrated ...odium chloride are very important:

- Chlorine is used to sterilise water.
- Hydrogen is used in the manufacture of margarine.
- Sodium hydroxide is also produced and is used in the production of soap.

## ...olten Sodium Chloride

...lid sodium chloride does not conduct electricity ...ecause the ions cannot move. If sodium chloride is ...eated until it melts, then electrolysis can occur.

...e electrolysis of molten sodium chloride produces ...dium and chlorine.

...loride ions, $Cl^-$, move to the positive electrode ...node) where chlorine is produced.

$$2Cl^- \rightarrow Cl_2 + 2e^-$$

...sitive sodium, $Na^+$, ions move to the negative ...ectrode (cathode) where sodium is produced.

$$Na^+ + e^- \rightarrow Na$$

## ...soluble Salts

...me salts are **insoluble**. They can be made by ...acting solutions of soluble salts.

...r example, insoluble barium sulfate is made by ...acting a solution of barium chloride with a ...lution of sodium sulfate. The reaction also produces ...dium chloride.

The barium sulfate is a **precipitate**.

| barium chloride | + | sodium sulfate | → | barium sulfate | + | sodium chloride |
|---|---|---|---|---|---|---|
| $BaCl_2(aq)$ | + | $Na_2SO_4(aq)$ | → | $BaSO_4(s)$ | + | $2NaCl(aq)$ |

- The state symbol (aq) means aqueous (it is dissolved in water).
- The state symbol (s) means solid.

### SUMMARY

- **Electrolysis of concentrated sodium chloride solution makes chlorine, hydrogen and sodium hydroxide.**
- **Electrolysis of molten sodium chloride produces sodium and chlorine.**
- **Some salts are insoluble.**

### QUESTIONS

#### QUICK TEST

1. Give one use of salts.

2. Name an insoluble salt.

3. Give the word equation for the reaction between barium chloride and sodium sulfate.

4. What does the state symbol (s) indicate?

5. What does the state symbol (aq) indicate?

#### EXAM PRACTICE

1. The compound lead bromide, $PbBr_2$, does not conduct electricity when solid but does conduct electricity when it is molten.

   Explain why, and predict the reactions that take place at the anode and the cathode.
   **(4 marks)**

# Copper

## Uses

Copper is a very useful metal. Copper is:

- a good electrical conductor
- a good thermal conductor
- unreactive.

Copper is widely used for electrical wiring and plumbing.

## Extraction

Copper is an unreactive metal. It can be extracted from its ore by heating. Copper produced in this way may contain some impurities.

Copper can be purified by electrolysis.

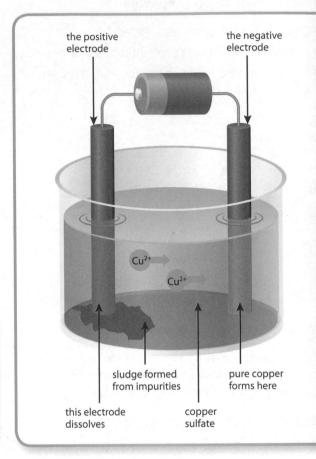

the positive electrode

the negative electrode

$Cu^{2+}$

$Cu^{2+}$

sludge formed from impurities

pure copper forms here

this electrode dissolves

copper sulfate

## Electrolysis

In electrolysis the impure copper is used as the positive electrode (anode). Pure copper is used as the negative electrode (cathode).

A solution that contains $Cu^{2+}$ ions is also used.

At the anode, copper atoms give up electrons to form copper ions. These ions dissolve into solution, so the anode loses mass.

$$Cu(s) \rightarrow Cu^{2+}(aq) + 2e^-$$

At the cathode, copper ions gain electrons to form copper atoms. Over time the cathode gets coated in very pure copper, so the cathode gains mass.

$$Cu^{2+}(aq) + 2e^- \rightarrow Cu(s)$$

- The loss in mass of the anode is equal to the gain mass of the cathode.
- The higher the electrical current, the more quickly the copper is produced at the cathode.
- The longer the time, the more copper is produced at the cathode.

The relationship between the charge transfer, current and charge is given by the following equation:

$$Charge = current \times time$$

## Electrolysis of Ionic Compounds

Ionic substances contain atoms or groups of atoms that have a charge. The ions in solid ionic substances have fixed positions but if the substances are dissolved or heated until they melt, the ions can mov

The electrolysis of molten **aluminium oxide**, $Al_2O_3$, produces aluminium and oxygen.

At the anode:

$$2O^{2-} \rightarrow O_2 + 4e^-$$

At the cathode:

$$Al^{3+} + 3e^- \rightarrow Al$$

The electrolysis of molten **lead bromide**, $PbBr_2$, produces lead and bromine.

At the anode:

$$2Br^- \rightarrow Br_2 + 2e^-$$

At the cathode:

$$Pb^{2+} + 2e^- \rightarrow Pb$$

The electrolysis of molten lead iodide, $PbI_2$, produces lead and iodine.

At the anode:

$$2I^- \rightarrow I_2 + 2e^-$$

At the cathode:

$$Pb^{2+} + 2e^- \rightarrow Pb$$

The electrolysis of molten **potassium chloride**, KCl, produces potassium and chlorine.

At the anode:

$$2Cl^- \rightarrow Cl_2 + 2e^-$$

At the cathode:

$$K^+ + e^- \rightarrow K$$

## SUMMARY

- Copper is a useful metal.
- Copper is extracted from its ore by heating and is then purified by electrolysis.
- Molten or dissolved ionic compounds can be separated by electrolysis.

## QUESTIONS

### QUICK TEST

1. How can copper be extracted from its ore?

2. How is copper purified?

3. What is the positive electrode called?

4. What is the negative electrode called?

5. During the purification of copper by electrolysis, what is the formula of the copper ion in this solution?

### EXAM PRACTICE

1. Copper can be purified by electrolysis. Impure copper is placed at the positive electrode (anode).

   a) Give the symbol equation for the reaction that takes place at the positive electrode.
   **(2 marks)**

   b) Copper atoms are deposited at the negative electrode. Give the symbol equation for the reaction that takes place at the negative electrode.
   **(2 marks)**

# Acids and Alkalis

## Strength of Acids and Alkalis

We describe the strength of an acid or an alkali by the extent to which it is ionised in water.

**Strong acids** are completely ionised in water. For example:

- hydrochloric acid
- sulfuric acid
- nitric acid

**Strong alkalis** are also completely ionised in water. For example:

- sodium hydroxide
- potassium hydroxide

**Weak acids** are only partially ionised in water. For example:

- ethanoic acid
- citric acid
- carbonic acid

**Weak alkalis** are also only partially ionised in water. For example:

- ammonia

The amount of ionisation affects conductivity.

- Hydrochloric acid is a strong acid. It completely ionises in water so it has a high electrical conductivity.

$$HCl \rightarrow H^+ + Cl^-$$

- Ethanoic acid is a weak acid. It only partially ionises in water so it has a much lower electrical conductivity.

$$CH_3COOH \rightleftharpoons CH_3COO^- + H^+$$

Electrolysis of both of these acids produces hydrogen gas at the negative electrode.

**Many Foods Contain Weak Acids**

COLA

**Many Cleaning Materials Contain Alkalis**

## Concentration

The concentration of an acid or an alkali is the number of moles in 1 dm³.

If the same concentration of hydrochloric acid and ethanoic acid are reacted with magnesium, the ethanoic acid reacts more slowly than the hydrochloric acid.

This is because the hydrochloric acid has a higher concentration of hydrogen ions, so there is a greater collision frequency between magnesium and the hydrogen ions and the reaction happens faster.

## Indicators

Indicators are special chemicals that are one colour in acidic conditions and another colour in alkaline conditions.

- Red litmus is red in acidic and neutral conditions and turns blue in alkaline conditions.

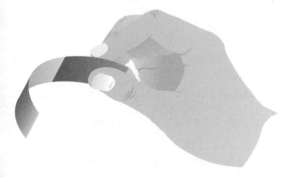

- Blue litmus is blue in neutral and alkaline conditions and turns red in acidic conditions.

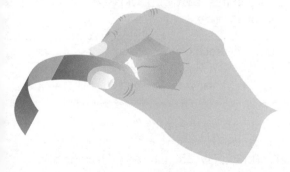

## SUMMARY

- Strong acids and alkalis are completely ionised in water.

- Weak acids and alkalis are partially ionised in water.

- The concentration of an acid or alkali is the number of moles in 1 dm³.

- Indicators are one colour in acidic conditions and another colour in alkaline conditions.

## QUESTIONS

### QUICK TEST

1. Why is citric acid known as a weak acid?

2. Write the formula of a hydroxide ion.

3. Why does hydrochloric acid have a higher electrical conductivity than ethanoic acid?

4. Name **three** strong acids.

5. Name a weak alkali.

### EXAM PRACTICE

1. In experiment A, 1 g of powdered calcium carbonate was reacted with 10 cm³ of 1 mol dm⁻³ sulfuric acid.

   In experiment B, 1 g of powdered calcium carbonate was reacted with 10 cm³ of 1 mol dm⁻³ ethanoic acid.

   Explain why bubbles were observed in both reactions and why the bubbles were produced more quickly in the first experiment. **(3 marks)**

# Making Salts

## Making Salts

Salts are very important compounds. They can be made by reacting acids with metals, metal carbonates, metal oxides or metal hydroxides.

● Hydrochloric acid forms **chloride** salts.
● Nitric acid forms **nitrate** salts.
● Sulfuric acid forms **sulfate** salts.

Sulfuric acid is used in the manufacture of fertilisers and in car batteries.

## Solutions

If a solution has a pH of 7 it is **neutral**. If the pH is less tha 7 it is an **acid** and if the pH is more than 7 it is an **alka**

An alkali is a soluble base.

Acids react with bases to form a salt and water.

The reaction between an acid and a base is called **neutralisation**.

Acidic solutions contain **hydrogen**, $H^+$, ions. Alkaline solutions contain **hydroxide**, $OH^-$, ions.

During **neutralisation** reactions, hydrogen ions react with hydroxide ions to form water, which is neutral.

$$H^+ (aq) + OH^- (aq) \rightarrow H_2O(l)$$

● The state symbol (l) means liquid.
● The state symbol (aq) means aqueous or dissolve in water.

## Metals

Fairly reactive metals react with acids to form a salt and hydrogen.

| magnesium + | hydrochloric acid | → | magnesium chloride | + hydrogen |
|---|---|---|---|---|
| Mg + | 2HCl | → | MgCl$_2$ + | H$_2$ |

### Metal Carbonates

Metal carbonates react with acids to form a salt, wate and carbon dioxide.

| calcium carbonate + | hydrochloric acid | → | calcium chloride | + water + | carbon dioxide |
|---|---|---|---|---|---|
| CaCO$_3$ + | 2HCl | → | CaCl$_2$ | + H$_2$O + | CO$_2$ |

The metal carbonate is added to the acid until the reaction stops (there is no more fizzing). The excess metal carbonate is then filtered off leaving a salt solution. If the salt solution is warmed carefully, the water evaporates leaving crystals of the salt.

## Metal Oxides

Metal oxides react with acids to form a salt and water.

copper oxide + nitric acid → copper nitrate + water

$$CuO + 2HNO_3 \rightarrow Cu(NO_3)_2 + H_2O$$

## Metal Hydroxides

Metal hydroxides also react with acids to form a salt and water.

potassium hydroxide + nitric acid → potassium nitrate + water

$$KOH + HNO_3 \rightarrow KNO_3 + H_2O$$

An indicator can be used to show when the acid and alkali have completely reacted.

## Ammonia

Ammonia can be reacted with acids to form ammonium salts.

ammonia + hydrochloric acid → ammonium chloride

$$NH_3 + HCl \rightarrow NH_4Cl$$

## Precipitation Reactions

A precipitate is a solid formed when two solutions react together. Precipitates are insoluble.

## SUMMARY

- In neutralisation reactions $H^+$ ions react with $OH^-$ ions to form water, $H_2O$.
- Metals react with acids to form a salt and hydrogen.
- Metal carbonates react with acids to make a salt, water and carbon dioxide.
- Metal oxides and metal hydroxides react with acids to form a salt and water.

## QUESTIONS

### QUICK TEST

1. If a solution has a pH of 7, what type of solution is it?

2. If a solution has a pH of 10, what type of solution is it?

3. If a solution has a pH of 2, what type of solution is it?

4. Name the salt made when magnesium reacts with hydrochloric acid.

5. Name the salt made when calcium carbonate reacts with hydrochloric acid.

### EXAM PRACTICE

1. Egg shells contain calcium carbonate. If hydrochloric acid is placed on an egg shell, the acid reacts with the calcium carbonate.

   a) Write a word equation to sum up this reaction.
   **(1 mark)**

   b) Suggest a value for the pH of hydrochloric acid.
   **(1 mark)**

# Metals

## Metallic Bonding

The atoms in a pure metal have a regular arrangement. This means that the atoms can slide over each other quite easily so metals can be bent into shape, or drawn into wires.

Metals can be mixed to form alloys. In alloys, the atoms have different sizes so it is harder for the layers of atoms to pass over each other. This makes alloys harder than pure metals.

In metals, the **electrons** in the outermost shell of an atom are **delocalised** and free to move throughout the whole structure. This means that metals consist of positive metal ions and negative delocalised electrons.

Metallic bonding is the attraction between these positive metal ions and the negative delocalised electrons. This is an **electrostatic** attraction.

Metals are good thermal and electrical conductors because the delocalised electrons are free to move throughout the structure.

Metals have high melting points because of the strong forces of attraction between the metal ions and the delocalised electrons.

## Metal Carbonates

When many metal **carbonates** are heated, they decompose to form a metal oxide and carbon dioxide. This is known as a thermal decomposition reaction. If carbon dioxide is bubbled through limewater it turns the limewater cloudy.

| copper carbonate | → | copper oxide | + | carbon dioxide |
|---|---|---|---|---|
| $CuCO_3$ | → | $CuO$ | + | $CO_2$ |
| iron(II) carbonate | → | iron oxide | + | carbon dioxide |
| $FeCO_3$ | → | $FeO$ | + | $CO_2$ |
| manganese carbonate | → | manganese oxide | + | carbon dioxide |
| $MnCO_3$ | → | $MnO$ | + | $CO_2$ |
| zinc carbonate | → | zinc oxide | + | carbon dioxide |
| $ZnCO_3$ | → | $ZnO$ | + | $CO_2$ |

We can tell that a chemical reaction has taken place when there is a change of colour, which shows us that a new substance has been formed.

## Hydroxide Tests

We can identify the metals present in metal salt solutions by adding **sodium hydroxide solution**.

If the metal ion forms a **precipitate**, we can use the colour of the precipitate to identify the metal present.

- copper(II) – pale blue precipitate
- iron(II) – green precipitate
- iron(III) – brown precipitate.

# ransition Metals

nsition metals are found in the middle of the riodic table.

ny transition metals and their compounds are eful catalysts.

n is used in the manufacture of ammonia, and :kel is used in the manufacture of margarine.

st iron is made into the alloy **steel**, which is used to ke cars and bridges because it is strong and cheap.

pper is used to make **brass**, which is used in electrical ring because it is a good electrical conductor.

Steel Bridge

# perconductors

me metals behave as **superconductors** at very low nperatures.

perconductors are special because when they duct electricity, they have very little or no resistance. e of superconductors will be limited until we can velop types that will work at room temperature.

## SUMMARY

- Metals have special properties because of metallic bonding.
- Transition metals are very widely used.
- When heated, many metal carbonates decompose.
- Hydroxide tests can be used to identify some metal ions.
- Superconductors have very little electrical resistance at low temperatures.

## QUESTIONS

### QUICK TEST

1. What is metallic bonding?

2. What are mixtures of metals called?

3. What is the catalyst used in the manufacture of ammonia?

4. What is the test for carbon dioxide?

### EXAM PRACTICE

1. Describe the structure of metals in terms of the ions and electrons involved. Explain how metallic structure affects the melting points and electrical conductivity of metals.

   *The quality of your written communication will be assessed in your answer to this question.*

   **(6 marks)**

# Useful Metals

## Aluminium

Pure aluminium has a low **density** but is too soft for many uses.

Aluminium can be mixed with other metals to form **alloys**. These alloys combine low density with high strength.

Other common alloys are:

- **amalgam** (mainly mercury)
- **brass** (copper and zinc)
- **solder** (lead and tin)
- **bronze** (copper and tin).

Aluminium is a reactive metal. It is extracted from its ore, bauxite, by **electrolysis**. This is an expensive process because it involves many steps and requires lots of energy.

Aluminium is actually much more reactive than it appears. This is because aluminium objects quickly react with oxygen to form a layer of aluminium oxide, which prevents any further reaction from occurring.

Aluminium can be used to make many things, including:

- drinks cans
- bicycles
- aeroplanes.

## Aluminium Cars

Car bodies are normally made from steel, but they can also be made from aluminium.

There are advantages to using aluminium to build cars

- The car body is lighter, so the car has better fuel economy.
- The car body corrodes less, so it may last for longer.

One of the disadvantages of using aluminium is that a aluminium car body is more expensive to produce.

## Protection of the Environment

There are several advantages of **recycling** old aluminium objects:

- Landfill sites are not filled up as quickly.
- Recycling old aluminium objects uses much less energy than extracting aluminium from its ore.
- Recycling old aluminium objects means that less aluminium ore needs to be extracted, which protects the environment.

## Titanium

Titanium is a very useful metal:

- It has a low density.
- It is strong.
- It is very resistant to corrosion.
- It has a very high melting point.

Titanium is a reactive metal, so extracting it from its ore (rutile), is a difficult process.

## Copper

Copper is a very useful metal:

- It is a good electrical conductor.
- It is a good thermal conductor.
- It is very resistant to corrosion.
- It is unreactive.

Copper is widely used for electrical wiring and plumbing. Traditionally, we extract copper from its ores and then purify it using electrolysis. Today we have to extract copper from ores that actually contain very little copper. This means that very large quantities of rock must be quarried, which causes environmental problems.

Scientists are developing ways to extract copper from low grade ores to try to minimise these environmental problems.

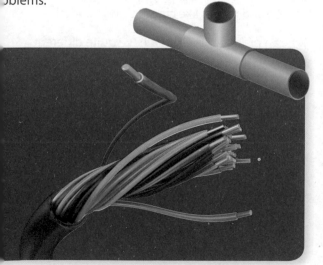

### SUMMARY

- An alloy is a mixture that contains at least one metal.

- Aluminium cars are expensive to produce but they do not corrode and are lighter than steel cars.

- Recycling materials protects our environment.

- Copper and titanium have special properties that make them very useful.

## QUESTIONS

### QUICK TEST

1. Why is pure aluminium not widely used?

2. What is a mixture of metals called?

3. What is brass made from?

4. Give one use of aluminium.

5. What is the ore of titanium called?

### EXAM PRACTICE

1. Car bodies can be made from steel or from aluminium. Explain how aluminium is extracted from its ore. What are the advantages and disadvantages of making a car body from aluminium?

   *The quality of your written communication will be assessed in your answer to this question.*

   **(6 marks)**

# Iron and Steel

## Extraction of Iron

An **ore** is a rock that contains metals in such high concentrations that it is economically worthwhile to extract the metal from the rock.

Iron is less reactive than carbon. To extract iron from iron(III) oxide, the oxygen must be removed. This reaction is called reduction.

Iron produced by a blast furnace contains quite large amounts of carbon. If it is allowed to cool down and solidify, **cast iron** is produced. Cast iron contains around 96% pure iron.

Cast iron is:

- very hard
- strong
- resistant to corrosion
- very brittle.

## Wrought Iron

**Wrought iron** is produced by removing the carbon from cast iron. Because it is very pure, the iron atoms form a very regular arrangement. This means that the layers of iron atoms can slip over each other very easily, which makes wrought iron very easy to shape. Unfortunately, it can be too soft for many uses.

## Steel

Most iron is made into **steel**. First the carbon impurities are removed, then carefully controlled amounts of carbon and metals like chromium are added.

We can produce steels that have quite different properties:

- Low carbon steels are soft and easy to shape.
- High carbon steels are hard but brittle.
- **Stainless steel** is very resistant to corrosion.

Steel is harder than wrought iron because it contains atoms of iron, carbon and other metals that are different sizes. This means the atoms cannot form a regular arrangement. This makes it very difficult for the layers to slip over each other, which makes steel hard.

Steel is also less likely to corrode than pure iron.

## Rusting

The corrosion of iron is called **rusting**. During rusting, iron reacts with water and oxygen to form hydrated iron(III) oxide.

This can be summarised by the following word equation:

iron + water + oxygen → hydrated iron(III) oxide

Salt increases the rate of rusting, so cars at the seaside may rust faster than cars away from the coast.

In many examples oxygen is lost in reduction reactions. During the extraction of iron from iron oxide, oxygen is removed from the iron oxide, so the iron oxide is reduced to form iron.

## SUMMARY

● Iron is extracted in a blast furnace.

● Cast iron contains about 4% carbon.

● Wrought iron is soft and easy to shape.

● Iron can be made into steel, which is less likely to rust than pure iron.

● The corrosion of iron is called rusting.

## QUESTIONS

### QUICK TEST

1. Name the iron compound found in iron ore.

2. What are the properties of a high carbon steel?

3. How is iron extracted from iron(III) oxide?

4. Iron is made by removing oxygen from iron oxide. What has happened to the iron oxide in this reaction?

5. Roughly, what is the percentage of iron in cast iron?

### EXAM PRACTICE

1. Use the words below to complete the table.                    (4 marks)

cast iron        wrought iron        low carbon steel        iron(III) oxide

| Name | Description |
|------|-------------|
| a) | The name of very pure iron |
| b) | An alloy containing iron and carbon, which is easy to shape |
| c) | A material produced when iron from a blast furnace is allowed to cool down and solidify |
| d) | The compound found in iron ore |

# Water

## Uses of Water

Water is a very important resource, which needs to be conserved.

Water is used as:

- a coolant
- a raw material to make new chemicals from
- a solvent.

## Water Pollution

Water can be polluted by:

- nitrates from fertiliser run-off
- lead compounds from lead pipes
- pesticides sprayed near to water resources.

## Nitrate Problems

Fertilisers are used to help plants to grow, but they can enter waterways and cause problems.

If fertiliser runs off into waterways, the nitrates in fertilisers can cause algae to grow at a fast rate. Eventually the algae die and bacteria begin to decompose them. The bacteria use up the oxygen in the water. Fish and other aquatic life cannot get enough oxygen and may die. This is called **eutrophication**.

In the UK, water is taken from areas that are well away from any sources of pollution, including:

- lakes
- rivers
- aquifers (rocks that contain water)
- reservoirs.

There is only a limited supply of fresh water, so it is important that we conserve water when we can.

# Water Purification

Clean, safe water is extremely important.

The following stages happen in the water purification process:

- Any large pieces of debris such as sticks are removed.
- The water is filtered to remove suspended particles such as clay.
- The water is passed through filter beds of charcoal and sand to remove unpleasant smells and make the water taste better.
- Finally, the water is chlorinated to reduce the number of microorganisms to acceptable levels.

Water can be obtained from seawater by distillation, but this takes a lot of energy and is only used where fuel is very cheaply available.

## Water Filters

The taste and quality of tap water can be improved by filtering the water. These filters contain carbon, silver and ion exchange resins.

## Insoluble Salts

Some salts are insoluble. They can be made by reacting solutions of soluble salts.

A solution of silver nitrate can be used to test for chloride, bromide or iodide ions. First dilute nitric acid is added and then the silver nitrate solution is added.

- If silver nitrate is added to a solution that contains chloride ions, a white precipitate of silver chloride is seen.
- If it is added to a solution that contains bromide ions, a cream precipitate of silver bromide is seen.
- If it is added to a solution that contains iodide ions, a pale yellow precipitate of silver iodide is seen.

## SUMMARY

- Water is collected from areas away from sources of pollution.
- It is then purified to make it safe to drink.
- Water filters can be used to improve water quality and taste.
- Silver nitrate solution can be used to test for chloride, bromide and iodide ions.

## QUESTIONS

### QUICK TEST

1. What is an aquifer?
2. During purification, why is water filtered?
3. Why is water chlorinated?
4. Why is distillation not widely used to purify water?
5. What compounds found in fertilisers can cause environmental problems?

### EXAM PRACTICE

1. Some silver salts are insoluble. Explain how chloride, bromide and iodide ions can be detected in water samples. Include the names of the silver compounds formed in these reactions.

   *The quality of your written communication will be assessed in your answer to this question.*

   **(6 marks)**

# Hard and Soft Water

## The Water Cycle

As the Sun warms the sea, some of the water evaporates to form **water vapour**. This condenses to form clouds.

The water droplets in the clouds join together to produce rain, which falls back to the Earth's surface. In this way, water is continually moved around the **water cycle**.

## Solubility

The **solubility** of a substance (solute) is measured as the number of grams of a substance that will dissolve in 100 g of water at a given temperature.

For solid substances, solubility generally increases as the temperature increases.

If a solution is **saturated**, then it cannot dissolve any more solute at that temperature. If the temperature of the solution is lowered, then some of the solute that had been dissolved will come out of solution.

## Gases

Many gases are soluble in water. The solubility of gases increases as the pressure increases and decreases as the temperature increases.

Carbonated drinks are made by dissolving **carbon dioxide** gas in water under high pressure. When the pressure is released, the gas comes out of solution and we see bubbles. A cold fizzy drink will contain more dissolved carbon dioxide than a warm fizzy drink.

The **oxygen** dissolved in water is essential for fish and other aquatic life. If the temperature of the water is increased, less oxygen is dissolved and aquatic life may perish.

## Soft Water

Soft water does not contain calcium or magnesium ions. It readily forms a **lather** (bubbles) with soap.

# Hard Water

Hard water contains calcium and magnesium ions. There are two types of hard water:

- Permanently hard water – this can be caused by dissolved calcium sulfate and cannot be removed by boiling.
- Temporarily hard water – this can be caused by dissolved calcium hydrogen carbonate and can be removed by boiling.

Water becomes hard when it comes into contact with rocks that contain calcium and magnesium ions. Hard water is good for you. It produces stronger bones and teeth and reduces the risk of developing heart disease.

## Problems with Hard Water

Hard water, however, can also cause problems. More soap is required to form a lather with hard water than is required with soft water. In addition, **limescale** can form when hard water is heated and this can damage heating systems and kettles. Limescale can be removed by weak acids.

Hard water can be made into soft water by removing the calcium and magnesium ions. This can be done by adding **washing soda** (sodium carbonate). This reacts with the calcium or magnesium ions to form a precipitate of calcium carbonate or magnesium carbonate.

Water can also be softened by passing it through an **ion exchange column**. In the column, the calcium and magnesium ions are replaced by hydrogen and sodium ions.

Both hard and soft water form a good lather with soapless detergents.

## SUMMARY

- Water is a good solvent for many substances.
- Soft water readily forms a lather with soap.
- Hard water contains calcium and magnesium ions.
- Hard water can be softened using an ion exchange column.

## QUESTIONS

### QUICK TEST

1. In the water cycle, why does water evaporate?

2. Name the gas that is dissolved in water to make fizzy drinks.

3. For solids, how does increasing the temperature affect solubility?

4. For gases, how does increasing the temperature affect solubility?

5. For gases, how does increasing the pressure affect solubility?

### EXAM PRACTICE

1. Fizzy drinks contain dissolved carbon dioxide.

   How does changing the temperature of the fizzy drink affect the amount of carbon dioxide that is dissolved in it?

   **(1 mark)**

# Detergents

## Washing Powders

We use washing powders to help get our clothes clean. They have lots of ingredients, including:

- **detergents** to remove the dirt from the fabric
- water softeners to remove the hardness from the water
- bleaches to remove coloured stains
- **enzymes** that help to remove stains at low temperatures; this helps consumers to save money and helps to protect the environment
- optical brighteners to make the clothes look really clean.

## Washing-up Liquids

We use washing-up liquids to help us to clean crockery and cutlery. They have lots of ingredients, including:

- a detergent to get the crockery and cutlery clean
- water to dilute the detergent so it is easier to use
- fragrance and colouring to make the washing-up liquids more attractive
- a water softener to make hard water softer
- a rinsing agent, which helps the water to drain away from the crockery/cutlery.

Water is a good **solvent** for most ionic compounds. If a compound **dissolves**, it is **soluble** and the mixture that is made is called a solution.

Grease and fats do not dissolve well in water. We can use detergents to remove them from surfaces.

Many detergents are salts made by a neutralisation reaction between an acid and an alkali.

Detergent molecules have two parts: a **hydrophilic** head group that is attracted to water molecules and a **hydrophobic** part that avoids water but is attracted to fat or grease. The detergent molecules surround the fat or grease stain, which can then be washed off.

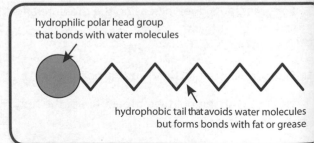

hydrophilic polar head group that bonds with water molecules

hydrophobic tail that avoids water molecules but forms bonds with fat or grease

## Washing Symbols

Washing symbols on clothes show the conditions that should be used to clean the garment.

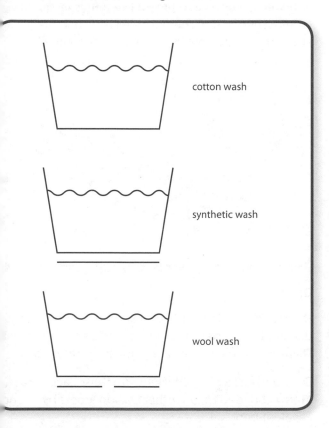

cotton wash

synthetic wash

wool wash

The temperature shown on the label is the maximum temperature that should be used. A lower temperature should be used but it may not get the clothes as clean.

## Dry Cleaning

Some fabrics can be damaged by washing them in water or may be stained with something that does not dissolve in water. These clothes can be **dry cleaned**.

In dry cleaning, the clothes are washed in a solvent other than water. This solvent is then extracted and the clothes are carefully pressed.

## Biological and Non-Biological Detergents

Biological detergents contain enzymes. Non-biological detergents do not contain enzymes.

## SUMMARY

- Washing powders and washing-up liquids contain many ingredients.

- Detergents help to dissolve fats and grease.

- Clothes can be cleaned by following the instructions found on washing labels, or by dry cleaning.

## QUESTIONS

### QUICK TEST

1. Why are bleaches added to washing powders?

2. Why are enzymes used in some washing powders?

3. Why is colouring and fragrance added to washing-up liquids?

4. Why is a rinsing agent added to washing-up liquids?

5. On a washing symbol, what does the number indicate?

### EXAM PRACTICE

1. Detergents are an important ingredient in washing powders.

   a) Why are detergents used?          **(1 mark)**

   b) Many detergents are salts. What should be added to an alkali to make a salt?    **(1 mark)**

   c) Name the end of the detergent molecule that is attracted to water molecules.
                                         **(1 mark)**

# Special Materials

## Smart Materials

Smart materials are very special materials that have one or more **property** that can be dramatically changed by changes in the environment.

**Thermochromic** pigments can be added to paints. If these paints are heated, the colour of the paint changes. Applications include designs on cups that appear when hot liquids are added.

**Nitinol** is another smart material. When a force is applied, nitinol stretches but when it is warmed up, it returns to its original shape.

## Nanoparticles

**Nanoparticles** consist of just a few hundred atoms, so they are incredibly small. Present uses of these materials include suncreams. Future uses could include better, smaller computers.

## Forms of Carbon

Diamond, graphite and fullerene are all **allotropes** (forms) of the element carbon.

### Buckminsterfullerene, $C_{60}$

Fullerenes consist of cages of carbon atoms held together by strong covalent bonds. These cages can be used to hold other molecules. Scientists are developing ways of delivering drugs using fullerenes. The most stable fullerene is called buckminsterfullerene. It has the formula $C_{60}$.

Buckminsterfullerene is a black solid.

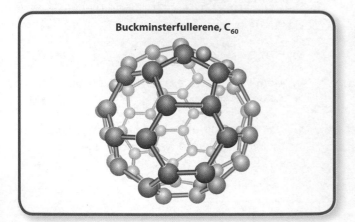
Buckminsterfullerene, $C_{60}$

## Nanotubes

Fullerenes can be made into **nanotubes**. Nanotubes are very hard and strong. They can be made into lightweight sports equipment like tennis racquets Other uses of nanotubes include industrial catalysts and semi-conductors for use in electrical circuits.

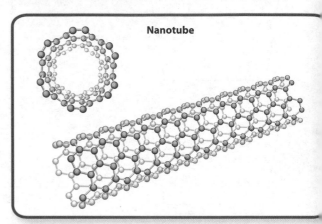
Nanotube

## Diamond

**Diamond** is a highly valued gemstone. High-quality diamonds are **lustrous**, colourless and are used in jewel Diamonds can also be used to make cutting tools.

Diamond has a giant covalent structure. Each carbon atom is bonded to four other carbon atoms by strong covalent bonds.

Diamond has a very high melting point and is hard because it has lots of strong bonds. It does not conduct electricity because there are no free electror or ions to move.

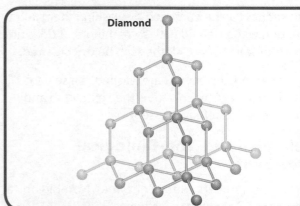
Diamond

### Graphite

**Graphite** is used in pencil leads, as a lubricant and to make electrodes. In graphite, each carbon atom is bonded to three other carbon atoms in the same layer by strong covalent bonds.

Graphite has a high melting point because it has lots of strong covalent bonds. There are much weaker forces of attraction between the layers. This means that the layers of carbon atoms can pass easily over each other.

Graphite conducts electricity because the electrons in the weaker bonds between layers are able to move.

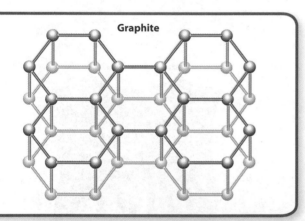

**Graphite**

## SUMMARY

- Smart materials have one or more property that alters due to changes in the environment.

- Nanoparticles contain a relatively small number of atoms.

- Buckminsterfullerene, $C_{60}$, nanotubes, diamond and graphite are all allotropes of carbon.

## QUESTIONS

### QUICK TEST

1. Give **one** use of graphite.

2. What is the formula of buckminsterfullerene?

3. Name **three** allotropes of carbon.

4. Why are diamonds used in jewellery?

5. What colour is solid buckminsterfullerene?

### EXAM PRACTICE

1. Diamonds have a very high melting point and do not conduct electricity.

   a) Describe the structure of diamond. **(2 marks)**

   b) Why does diamond have a high melting point? **(2 marks)**

   c) Why does diamond not conduct electricity? **(1 mark)**

# New Materials

## Gore-tex®

Gore-tex is used to make waterproof objects like jackets. It consists of a thin membrane, which is used to coat fabrics. The membrane has lots of little holes. Liquid water is too big to go through these holes, so the fabric is waterproof. Water vapour is small enough to pass through the holes, so it is **breathable**.

In the past, waterproof jackets were made from nylon. This material was lightweight, hard wearing and waterproof, but because it was not breathable, perspiration could make a nylon coat quite uncomfortable to wear.

## PTFE

PTFE is better known by its trade name of Teflon®. The properties of PTFE are as follows:

- It is very unreactive.
- It has a very slippery surface.

PTFE was first discovered accidentally by scientists investigating refrigerant gases. Today it is used to make non-stick saucepans.

## Kevlar®

Kevlar is:

- lightweight
- very strong
- flexible.

Kevlar is used to make bulletproof vests.

## Lycra®

Lycra is a very **stretchy** material. It can be mixed with other fibres to make fabrics that can be used to make swimsuits.

## Thinsulate™

Thinsulate is an insulating material. It contains very small **fibres**. These fibres trap air that acts as a layer of insulation stopping body heat from escaping, so it keeps you warm.

Gore-tex Jacket

Bulletproof Vest

Lycra Swimwear

## Carbon Fibres

Carbon fibres:

- have a low density
- do not stretch
- are not compressed.

Carbon fibres are mixed with epoxy to make sports equipment such as squash racquets.

## Phosphorescent Pigments

Phosphorescent pigments absorb light energy and then release it over a period of time. They can be used to make objects that glow in the dark, such as wristwatch dials.

## Synthesis

Some new chemicals, such as medicines, are made in batch processes. These chemicals are made in relatively small amounts when they are needed. Other new chemicals, such as sulfuric acid, are made all the time in continuous processes. These chemicals are needed in large amounts.

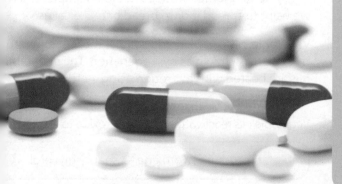

## SUMMARY

- Gore-tex is used to make objects that are waterproof and breathable.

- Scientists have produced many new materials that have improved people's lives.

- New chemicals can be made in batch processes or in continuous processes.

## QUESTIONS

### QUICK TEST

1. Why can liquid water not pass through Gore-tex ?

2. Which material was widely used to make waterproof jackets in the past?

3. What is the trade name for PTFE?

4. How is PTFE used?

5. What is Kevlar used to make?

### EXAM PRACTICE

1. Thinsulate is a very useful material.

   a) Describe the structure of Thinsulate.
      **(1 mark)**

   b) Explain why Thinsulate might be used to make a hat. **(3 marks)**

# Vegetable Oils

## Oils as Foods

Vegetable oils are important foods. They are a good source of **energy** and of the vitamins A and D.

Popular vegetable oils include sunflower oil and olive oil.

Vegetable oils can also be used as **biofuels**. When they are burned, they release lots of energy.

Vegetable oils can be produced from:

- seeds
- nuts
- fruits.

The oil is removed by first crushing up the plant material and then collecting the oil.

Vegetable oils are unsaturated because they contain carbon double bonds. We can check this by adding bromine water. The bromine water changes colour from brown to colourless.

Some vegetable oils contain many carbon double bonds. These are described as **polyunsaturated** fats. Doctors believe that polyunsaturated fats are good for our health.

**Olive Oil**

## Emulsions

An **emulsion** is a mixture of two immiscible liquids. Immiscible liquids are ones that do not blend together.

Salad dressing is an example of an everyday emulsion. Salad dressings are made by shaking oil with vinegar so that the two liquids mix together, but the oil and vinegar soon separate out.

To stop this from happening we can add an **emulsifier**. One end of an emulsifier molecule is attracted to the water in vinegar (hydrophilic), while the other end is attracted to the oil molecules (hydrophobic).

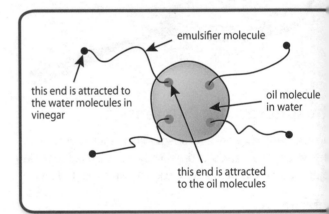

emulsifier molecule

this end is attracted to the water molecules in vinegar

oil molecule in water

this end is attracted to the oil molecules

Milk is an oil-in-water emulsion.

Butter is a water-in-oil emulsion.

## Hydrogenated Vegetable Oils

Most vegetable oils are liquids at room temperature. Liquid oils can be very useful, but there are times when we might prefer a solid fat, for example, when we want to spread it onto bread or make cakes.

Vegetable oils can be made solid at room temperature by a process known as hydrogenation. The vegetable oils are heated with hydrogen using a nickel catalyst around 60 °C.

Vegetable oils react with sodium hydroxide to form soap.

fat + sodium hydroxide → soap + glycerol

## aint

aint can be used to make something look more tractive, or to protect a surface.

aint is a special type of mixture called a **colloid**. It onsists of tiny coloured particles that are suspended a solvent. Types of paint include oil paints and mulsion paints.

hermochromic pigments change colour as the mperature changes.

hosphorescent pigments take in energy and then lease it later as light.

### SUMMARY

● Vegetable oils can be produced from seeds, nuts and fruits.

● Emulsions are mixtures of two immiscible liquids.

● Vegetable oils can be solidified by hydrogenation.

● Paints are colloid mixtures.

## QUESTIONS

### QUICK TEST

1. Which vitamins are found in vegetable oils?

2. Name **two** types of vegetable oil.

3. Which parts of plants can we extract vegetable oils from?

4. Why can vegetable oils be described as 'unsaturated'?

5. What do salad dressings contain?

### EXAM PRACTICE

1. Use the words below to complete the table.                    (4 marks)

emulsion       bromine water       biofuel       polyunsaturated

| Name | Description |
|---|---|
| a) | A mixture of two immiscible liquids |
| b) | A fat with many carbon double bonds |
| c) | A material made from living organisms that can be burned to release energy |
| d) | A chemical used to test for a carbon double bond |

# Food Additives

## Cooking Food

Many foods are cooked before we eat them. The cooking of food is a chemical change.

Eggs and meat are good sources of **protein**. Potatoes are a good source of **carbohydrate**.

When we cook eggs or meat, the shape of the protein molecules are changed. Cooking potatoes makes them easier to digest.

So, we cook food to make it:

- taste better
- look better
- easier to digest.

The high temperatures used in cooking can also kill microorganisms in the food.

## Food Additives

Chemicals are added to foods to make them:

- look more attractive
- taste better
- have a longer shelf-life.

Chemicals that have passed safety tests and are approved for use throughout the European Union are called **E-numbers**.

Food additives can be natural substances or artificial substances.

Chemicals commonly added to foods include the following:

- **Emulsifiers** – used to keep unblendable liquids mixed together.
- **Colours** – used to make food look more attractive (artificial colourings can be detected by chromatography).
- **Flavours** – used to make food taste better.
- **Artificial sweeteners** – used to reduce the amou of sugar needed.
- **Preservatives** – used to increase the shelf-life of food by stopping harmful microorganisms from growing.
- **Antioxidants** – used to increase the shelf-life of foods that contain oils and fats by preventing reactions with oxygen.

## ublic Concern

he number of additives in food has caused some
ublic concern. However, scientists believe that
long as we eat a balanced diet including lots of
fferent foods, the permitted food additives are safe
r the vast majority of people.

## aking Powder

aking powder is used to make biscuits and cakes.
contains **sodium hydrogencarbonate**, $NaHCO_3$.

hen sodium hydrogencarbonate is heated, a thermal
ecomposition reaction takes place, producing
dium carbonate, water and carbon dioxide.

$$\text{sodium hydrogencarbonate} \rightarrow \text{sodium carbonate} + \text{water} + \text{carbon dioxide}$$

$$2NaHCO_3 \rightarrow Na_2CO_3 + H_2O + CO_2$$

e products include carbon dioxide gas, which
comes trapped in the dough mixture and makes
e dough rise. This gives cakes and biscuits a
ht texture.

## SUMMARY

- Cooking food is an example of a chemical reaction.

- E-numbers are used to improve foods, but some people are concerned about their use.

- When sodium hydrogencarbonate is heated, it produces sodium carbonate, water and carbon dioxide.

## QUESTIONS

### QUICK TEST

1. Where are E-numbers approved for use?

2. What does an emulsifier do?

3. Is a strawberry an artificial or a natural substance?

4. What type of additive can be added to a food to make it look better?

5. What type of food additive is added to oils and fats to stop them from reacting with oxygen?

### EXAM PRACTICE

1. What is the name of the chemical found in baking powder? **(1 mark)**

2. Why is baking power used when we make cakes or biscuits? **(1 mark)**

3. Complete the word equation to show the products of the thermal decomposition of sodium hydrogencarbonate:

   sodium hydrogencarbonate → sodium carbonate + water + _____ **(1 mark)**

# Crude Oil

## Formation of Fossil Fuels

Fossil fuels such as coal, oil and natural gas are formed over millions of years from the fossilised remains of dead plants and animals.

Fossil fuels are non-renewable. They take millions of years to form. Our reserves of fossil fuels are being used up very quickly.

## Fractional Distillation

Crude oil is a mixture. The most important compounds in crude oil are called **hydrocarbons**.

Hydrocarbons are compounds that only contain carbon and hydrogen atoms.

Short hydrocarbon molecules are valuable **fuels** for several reasons:

- They are easy to ignite.
- They have low boiling points.
- They are runny.

Longer hydrocarbon molecules are less valuable, but they are still widely used. Crude oil is separated into **fractions** (groups of molecules with a similar number of carbon atoms) by fractional distillation.

During **fractional distillation** crude oil is heated up until it evaporates. Short hydrocarbon molecules have low boiling points and reach the top of the fractionating column before they condense and are collected.

Longer hydrocarbon molecules have higher boiling points. They condense and are collected lower down the fractionating column.

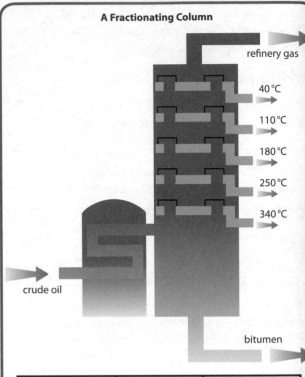

A Fractionating Column

refinery gas
40 °C
110 °C
180 °C
250 °C
340 °C

crude oil

bitumen

| No. carbon atoms in hydrogen chain | Temperature | Fraction collected |
|---|---|---|
| 3 | Less than 40 °C | Refinery gas |
| 8 | 40 °C | Petrol |
| 10 | 110 °C | Naphtha |
| 15 | 180 °C | Kerosene |
| 20 | 250 °C | Diesel |
| 35 | 340 °C | Oil |
| 50+ | Above 340°C | Bitumen |

## Uses

| Fraction | Use |
|---|---|
| Gases | Heating |
| Petrol | Fuel for vehicles |
| Naphtha | To make new chemicals |
| Kerosene | Jet fuel |
| Diesel and oil | Heating and as fuel for vehicles |
| Bitumen | To make roads |

## Cracking

Fractional distillation of crude oil produces large amounts of long hydrocarbon molecules.

These long molecules can be broken down into smaller, more useful and more valuable molecules by **cracking**. In this process, large molecules are heated until they evaporate and are then passed over a catalyst.

For example, decane is cracked to produce octane and ethene.

Decane $C_{10}H_{22}$ (from the naphtha fraction)

H–C–C–C–C–C–C–C–C–C–C–H (with H atoms above and below each C)

↓

Octane $C_8H_{18}$

H–C–C–C–C–C–C–C–C–H (with H atoms above and below each C)

+

Ethene $C_2H_4$

$C = C$ (with H atoms)

## SUMMARY

- Coal, oil and natural gas are fossil fuels.
- Crude oil is separated into its components by fractional distillation.
- Cracking turns long hydrocarbon molecules into smaller, more valuable molecules.

## QUESTIONS

### QUICK TEST

1. What is a hydrocarbon?

2. What is a fraction?

3. In a fractionating column, where are the shortest hydrocarbon molecules collected?

4. What is the name of the fraction that is used to make roads?

5. Which fraction is used for jet fuel?

### EXAM PRACTICE

1. Use the words below to complete the sentences.

   condense          evaporates
   fractional distillation     fractions

   Crude oil is separated by ___**a)**___ . The crude oil is heated until it ___**b)**___ . Short hydrocarbon molecules reach the top of the column before they ___**c)**___ and are collected. Groups of molecules with a similar number of carbon atoms are called ___**d)**___ .

   (4 marks)

# Fuels

## Fuels

A fuel is a substance that can be burned to release **energy**. This means the burning of a fuel is an exothermic reaction.

A good fuel should:

- be easy to ignite
- be widely available
- release lots of energy when it is burned
- burn cleanly without producing lots of soot
- not damage the environment.

Most fuels contain carbon and/or hydrogen.

**Hydrocarbons** are compounds that only contain carbon and hydrogen atoms. Petrol, diesel and fuel oil are all hydrocarbons. Coal is mainly made of carbon.

## Combustion

Complete combustion of hydrocarbon fuels produces **carbon dioxide**, $CO_2$, and **water vapour**, $H_2O$.

Scientists believe that carbon dioxide contributes to global warming.

Incomplete combustion of hydrocarbon fuels can occur if there is a limited supply of oxygen. This can occur in faulty gas appliances. The products of incomplete combustion include **carbon monoxide**, **CO**, and carbon.

Carbon monoxide is a toxic gas. It lowers the ability of red blood cells to carry oxygen around the body.

Unburnt carbon makes a flame look yellow and it is deposited as **soot**, making surfaces dirty.

## Hydrogen

Hydrogen is widely used as a fuel in space rockets, but it is not currently used as a fuel for cars.

| Advantages of using hydrogen as a fuel | Disadvantages of using hydrogen as a fuel |
|---|---|
| Releases lots of energy when it is burned. | It is a gas that can only be liquefied at very low temperatures or very high pressures. |
| Burns very cleanly. | It is difficult to store – a car would require a large heavy container to store the hydrogen. |
| The only product of combustion is water vapour, which does not harm the environment. | Hydrogen is not found naturally on Earth and would have to be made before being used. This would probably require the use of fossil fuels. |

**Sooty Fireplace**

# Biofuels

**Biofuels** are fuels that are produced from living things. Wood and ethanol are examples of biofuels.

## SUMMARY

● **Fuels are burned to release heat energy.**

● **Complete combustion of hydrocarbons produces carbon dioxide and water vapour.**

● **Incomplete combustion of hydrocarbons produces unburned carbon and carbon monoxide.**

● **Hydrogen can be used as a fuel.**

● **Biofuels are made from living things.**

## QUESTIONS

### QUICK TEST

1. Which **two** elements are found in hydrocarbon fuels?

2. What are the products of the complete combustion of hydrocarbon fuels?

3. Name the main element found in coal.

4. What is the environmental problem associated with carbon dioxide?

5. Why is the incomplete combustion of hydrocarbon fuels undesirable?

### EXAM PRACTICE

1. Complete the word equation below to show the product of the complete combustion of carbon. **(1 mark)**

   carbon + oxygen ⟶ _____

2. Carbon can also be burned in a limited supply of oxygen. Name one of the products of the incomplete combustion of carbon. **(1 mark)**

3. Many fuels also contain atoms of hydrogen. Complete the word equation to show the product of the combustion of hydrogen. **(1 mark)**

   hydrogen + oxygen ⟶ _____

## Carbon Cycle

Until recent times, the level of carbon dioxide in the atmosphere has remained fairly constant for a long time.

This is because of the carbon cycle, which moves carbon between the atmosphere, rocks and oceans.

# Air and Air Pollution

## Evolution of the Atmosphere

Today's atmosphere is composed of about 78% nitrogen, about 21% oxygen, about 1% argon and small amounts of other gases including carbon dioxide and water vapour. We can obtain nitrogen and oxygen from liquid air by fractional distillation.

During the first billion years of the Earth's life, there were enormous amounts of volcanic activity. The volcanoes released water vapour, which condensed to form the early **oceans**, while the other gases that were released formed the Earth's early atmosphere.

This atmosphere mainly consisted of **carbon dioxide**, like the present-day atmosphere of Mars and Venus.

During the next two billion years, **plants** evolved and began to cover the Earth's surface. These plants steadily removed carbon dioxide and produced oxygen by photosynthesis.

Ammonia in the Earth's early atmosphere reacted with oxygen to produce nitrogen. Nitrogen was also produced by denitrifying bacteria.

As oxygen levels increased an ozone layer, $O_3$, formed. The ozone layer filtered out harmful ultraviolet radiation so that more complex life forms could evolve.

In recent times, people have started to burn large amounts of fossil fuels. This has released large amounts of carbon dioxide back into the atmosphere.

## Acid Rain

Many fossil fuels contain small amounts of **sulfur**. When these fuels are burned, this sulfur reacts with oxygen to form sulfur dioxide. If this gas is released into the atmosphere, it can react with rainwater to form acid rain.

We can reduce the amount of acid rain produced by:

- burning fewer fossil fuels
- removing sulfur compounds from oil and natural gas
- removing sulfur dioxide from the waste gases in coal-powered power stations before they are released into the atmosphere.

acid rain

## Global Dimming

**Global dimming** is caused by smoke particles being released into the atmosphere when fuels are burned.

## Global Warming

Scientists have found that the temperature of the Earth seems to be gradually increasing.

Carbon dioxide, which is produced when fossil fuels are burned, is thought to be contributing to this effect, known as **global warming**.

The carbon dioxide traps heat energy reaching the Earth from the Sun. If global warming continues, the polar icecaps may eventually melt. This could cause huge environmental problems.

## Nitrogen Oxides

At the high temperatures inside car engines, nitrogen in the atmosphere may react with oxygen to form **nitrogen monoxide**. The nitrogen monoxide is then oxidised to form **nitrogen dioxide**. Nitrogen monoxide and nitrogen dioxide are known as nitrogen oxides, or $NO_x$.

## Reducing Pollution from Cars

The amount of atmospheric pollution caused by cars can be reduced by:

- using more efficient engines
- using public transport more
- using catalytic converters, which convert carbon monoxide to carbon dioxide, and nitrogen monoxide to nitrogen and oxygen.

---

## QUESTIONS

### QUICK TEST

1. What is the main gas in today's atmosphere?

2. What was the main gas in the Earth's early atmosphere?

3. Why did the amount of carbon dioxide in the Earth's early atmosphere decrease?

4. Name the gas that makes up around 20% of today's atmosphere.

5. Name the gas produced when sulfur reacts with oxygen.

### EXAM PRACTICE

1. The Earth's early atmosphere mainly consisted of carbon dioxide.

   a) Which planet has an atmosphere similar to the Earth's early atmosphere? **(1 mark)**

   b) Explain why the Earth's atmosphere today has less carbon dioxide and more oxygen than it has had in the past. **(2 marks)**

   c) What is the environmental problem associated with increased levels of carbon dioxide in the atmosphere? **(1 mark)**

---

## SUMMARY

- **In the past, the Earth's atmosphere mainly consisted of carbon dioxide.**

- **Acid rain can be formed when fuels that contain sulfur are burned.**

- **Scientists believe that global dimming and global warming are occurring.**

- **Catalytic converters help to reduce the amount of nitrogen monoxide produced by car engines.**

# Pollution

## Limestone

Limestone is an important raw material. It can be used as a building material and can be made into other important materials like cement and glass.

The advantages of limestone quarrying include the economic and social benefits of new jobs for local people.

The disadvantages of a limestone quarry include noise and dust from the quarry, and problems created by lorries transporting the limestone.

## Plastics

Plastics are a type of **polymer**. The properties of plastics depend on how they are made and what they are made from.

Polymers are very widely used and new applications are being developed including new '**intelligent**' **packaging**.

'Intelligent' packaging is used to improve the quality and safety of foods. For example, it can be used to remove water from inside a packet so that it is more difficult for bacteria or mould to grow and the food will stay fresh for longer.

## Problems with Plastics

Plastics are very useful. Most plastics, however, are **non-biodegradable**. This means that when plastic objects are thrown away, they remain in the environment. This can cause problems:

- Landfill sites fill up more quickly.
- If we try to get rid of the plastics by burning them, toxic gases may be produced.
- It is very difficult to recycle plastics because it is hard to separate them.

Because of these problems, scientists have developed a range of plastics that are biodegradable.

## Recycling

Recycling waste materials, including glass, metal and paper, helps to protect the environment because fewer raw materials are required. This also means that less waste needs to be disposed of.

This plastic bag is 100% degradable* but you can still reuse it!

*From date of manufacture, the plastic will start to degrade in approx. 18 months time. The whole process will take about 3 years. See bottom of bag for date of manufacture.

epi

## Problems with Oil Exploitation

Crude oil is a very important resource. It is found in porous rocks in the Earth's crust.

Oil is moved around the world in giant oil tankers. If an accident happens as the oil is being collected or transported, then the oil can escape and form an oil slick.

These slicks can harm animal and plant life and devastate beaches.

## Sustainable Development

Sustainable development balances the need for **economic development** with a respect for the **environment**, so that people can enjoy a good standard of living today without compromising the needs of future generations.

### SUMMARY

- Limestone quarries can bring advantages and disadvantages to local people.
- Plastics are useful but non-biodegradable plastics cause environmental problems.
- Oil slicks can harm animal and plant life.
- Sustainable development balances the need for economic growth with the protection of the environment.

### QUESTIONS

#### QUICK TEST

1. How can a new limestone quarry benefit people?

2. What are the disadvantages of having a limestone quarry nearby?

3. Name a type of polymer.

4. What could be a disadvantage of burning plastics?

5. Why is it difficult to recycle plastics?

#### EXAM PRACTICE

1. Scientists have developed 'intelligent' packaging for food.

    a) What can 'intelligent' packaging do?
    **(1 mark)**

    b) Why is this an advantage to consumers?
    **(2 marks)**

# Alkanes and Alkenes

## Alkanes

Carbon atoms form four bonds with other atoms, while hydrogen atoms only form one bond.

The **alkanes** are a family of **hydrocarbon** molecules. Hydrocarbon molecules only contain carbon and hydrogen atoms. Alkanes are saturated hydrocarbons because they do not contain carbon double bonds. Short alkane molecules are useful fuels.

Alkanes have the general formula $C_nH_{2n+2}$.

| Name | Chemical formula | Structure |
|------|------------------|-----------|
| Methane | $CH_4$ | H–C–H with H above and below — covalent bonds |
| Ethane | $C_2H_6$ | H–C–C–H with H atoms |
| Propane | $C_3H_8$ | H–C–C–C–H with H atoms |
| Butane | $C_4H_{10}$ | H–C–C–C–C–H with H atoms |

## Alkenes

The **alkenes** are another family of hydrocarbon molecules.

Alkenes are produced during cracking.

Alkenes are unsaturated hydrocarbons because they contain carbon double bonds. Alkene molecules are more reactive than alkane molecules.

They can be used to make new chemicals, including plastics.

Alkenes have the general formula $C_nH_{2n}$.

| Name | Chemical formula | Structure |
|------|------------------|-----------|
| Ethene | $C_2H_4$ | C=C with H atoms — double carbon bond |
| Propene | $C_3H_6$ | C=C–C with H atoms |

## Industrial Alcohol

Ethanol can be made from non-renewable sources. Ethanol produced in this way is called industrial alcohol. Ethene, which is produced by the cracking of long chain hydrocarbons, is reacted with steam to produce ethanol.

ethene + steam → ethanol

$C_2H_4$ + $H_2O$ → $C_2H_5OH$

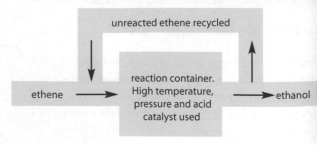

unreacted ethene recycled

ethene → reaction container. High temperature, pressure and acid catalyst used → ethanol

## Uses of Alcohol

Ethanol can be used as a fuel. It burns very cleanly, producing very little carbon monoxide. Ethanol can also be made by fermentation. During fermentation, glucose reacts to produce ethanol and carbon dioxide.

The glucose can be obtained from sugar cane or sugar beet. Ethanol produced in this way is a renewable fuel.

**Sugar Cane Plantation**

### SUMMARY

● **Alkanes are saturated hydrocarbons.**

● **Alkenes are unsaturated hydrocarbons.**

● **Industrial alcohol is produced from ethene and steam.**

● **Ethanol is a useful fuel.**

## QUESTIONS

### QUICK TEST

1. What is the general formula of an alkane?

2. Give a use for a short alkane molecule.

3. What do the lines in structural diagrams represent?

4. How can alkenes be made?

5. What is the formula of ethene?

### EXAM PRACTICE

1. Ethanol can be made by reacting ethene with steam.

   a) To which family does ethene belong to?
      **(1 mark)**

   b) Ethene is 'unsaturated' – what does this mean? **(1 mark)**

   c) Write a word equation for the reaction of ethene with steam. **(1 mark)**

# Alcohols, Acids and Esters

## Alcohols

Alcohols are a family of the organic compounds with the functional group hydroxyl, OH.

Short carbon chain alcohols are miscible with water (mix fully with water). The longer the carbon chain, the less soluble the alcohol.

Alcohols are neutral compounds (they have a pH of 7). They are widely used as solvents.

### Methanol

Methanol is the first member of the alcohol family.

$$H-\overset{\displaystyle H}{\underset{\displaystyle H}{C}}-O-H$$

### Ethanol

Ethanol (often just called alcohol) is also used as a fuel and is found in alcoholic drinks.

$$H-\overset{\displaystyle H}{\underset{\displaystyle H}{C}}-\overset{\displaystyle H}{\underset{\displaystyle H}{C}}-O-H$$

Ethanol can be oxidised by microbes or by chemical oxidising agents to form ethanoic acid.

## Carboxylic Acids

Carboxylic acids are another group of organic compounds. They have the carboxyl, COOH, functional group.

Examples include methanoic acid and ethanoic acid.

**Methanoic acid**

$$H-C\overset{\displaystyle O}{\underset{\displaystyle O-H}{}}$$

**Ethanoic acid**

$$H-\overset{\displaystyle H}{\underset{\displaystyle H}{C}}-C\overset{\displaystyle O}{\underset{\displaystyle O-H}{}}$$

Carboxylic acids have a sharp, distinctive odour. They are all weak acids and react with bases to form salts. Solutions of carboxylic acids have a pH less than 7.

Vinegar is an aqueous solution that contains ethanoic acid.

# sters

sters can be made by reacting alcohols with arboxylic acids.

$$\text{ethanol} + \text{ethanoic acid} \rightleftharpoons \text{ethyl ethanoate} + \text{water}$$

$$CH_3CH_2OH + CH_3COOH \rightleftharpoons CH_3COOCH_2CH_3 + H_2O$$

**Ethyl Ethanoate**

ters have pleasant, sweet odours and in nature are sponsible for the taste and smell of many fruits.

nthetic esters are widely used as flavourings in ocolates, sweets and other processed foods. They e also used in cosmetics like body sprays and as lvents in perfumes and aftershaves.

## SUMMARY

- Alcohols contain the hydroxyl, OH, group.
- Carboxylic acids contain the carboxyl, COOH, group.
- Esters are made by reacting alcohols and carboxylic acids together.

## QUESTIONS

### QUICK TEST

1. To what family does ethanol belong?

2. To what family does ethyl ethanoate belong?

3. To what family does methanol belong?

4. To what family does methanoic acid belong?

5. Suggest a pH for ethanoic acid.

### EXAM PRACTICE

1. Ethanol is a type of alcohol. It has the molecular formula $CH_3CH_2OH$.

   a) A sample of ethanol is tested using a pH meter. What do you expect the pH to be?
   **(1 mark)**

   b) Give a use of ethanol. **(1 mark)**

   c) Name the useful chemical produced when ethanol is oxidised by microbes. **(1 mark)**

# Polymers

## Polymerisation

In **addition polymerisation**, many small molecules are joined together to form a bigger molecule.

The small molecules are called **monomers**. These monomers are unsaturated molecules. The bigger molecule is called a polymer. Plastics are **polymers**.

This diagram represents the reaction between lots of ethene molecules to form polythene.

$$n \quad \begin{matrix} H & H \\ | & | \\ C = C \\ | & | \\ H & H \end{matrix} \rightarrow \left( \begin{matrix} H & H \\ | & | \\ C - C \\ | & | \\ H & H \end{matrix} \right)_n$$

This diagram represents the reaction between lots of propene molecules to form polypropene.

$$n \quad \begin{matrix} CH_3 & H \\ | & | \\ C = C \\ | & | \\ H & H \end{matrix} \rightarrow \left( \begin{matrix} CH_3 & H \\ | & | \\ C - C \\ | & | \\ H & H \end{matrix} \right)_n$$

Polymerisation reactions may involve other monomer units.

Polypropene is made by an addition polymerisation reaction between many propene molecules.

Polytetrafluoroethene (PTFE) is made by an addition polymerisation reaction between many tetrafluoroethene molecules.

The exact properties of the polymer made depends on:

- the length of the polymer chains
- the monomer used
- the conditions under which the polymer chain was made.

## Properties of Plastics

The structure and bonding within a material affects its properties.

The stronger the forces between the particles in a solid the higher the temperature at which the solid melts.

Modifying the structure of a polymer can influence its properties.

- Adding a **plasticiser** can make a plastic more flexible
- Lengthening the polymer chain increases the forces of attraction between the molecules and increases the melting point of the polymer.
- Making the polymer chains more aligned increases the forces of attraction between the molecules and increases the melting point.
- Some plastics consist of long polymer chains with very little cross linking between the chains. When these plastics are heated, the chains untangle and the plastic softens. This means that these plastics can be reshaped many times.
- Other plastics consist of polymer chains that are heavily **cross linked**. These polymers must be shaped when they are first made. When they are heated again they will not soften but may eventually burn.

Plastics are useful, but disposing of them can be difficult

# Useful Polymers

Polythene is cheap and strong. It is used to make plastic bags and bottles.

PVC is lightweight and rigid. It is used to make rainpipes and garden fences.

Polystyrene is cheap and lightweight. It is used to make packaging and plastic casing.

## SUMMARY

- Many small monomers join together to form polymers.

- Plastics are polymers.

- The structure and bonding within polymers affects their properties.

- Different plastics have different properties and therefore different uses.

## QUESTIONS

### QUICK TEST

1. Name the small molecules used to make polymers.

2. Name the polymer made from ethene.

3. Name the polymer made from propene.

4. How does adding a plasticiser affect a plastic?

5. What happens when a heavily cross-linked polymer is reheated?

### EXAM PRACTICE

1. Name the monomer used to make polythene. **(1 mark)**

2. Suggest a use for polythene. **(1 mark)**

3. How does lengthening the polymer chains in a plastic affect its melting point? Explain your answer. **(2 marks)**

# Limestone

## Heating Limestone

Limestone is a type of sedimentary rock. It is mainly made of **calcium carbonate**, $CaCO_3$.

When limestone is heated, it decomposes to form calcium oxide and carbon dioxide:

> calcium carbonate → calcium oxide + carbon dioxide

This is an example of a **thermal decomposition** reaction.

Calcium oxide is also known as **quicklime**. Quicklime can be reacted with water to form calcium hydroxide:

> calcium oxide + water → calcium hydroxide

Calcium hydroxide is also known as **slaked lime**.

**Limestone Quarry**

## Other Metal Carbonates

Other metal carbonates react in a similar way when they are heated, but in Group 1 only lithium carbonate decomposes at Bunsen temperatures.

## More Useful Materials

Other useful materials made from limestone include:

- cement – made by heating powdered limestone and powdered clay and then adding water
- mortar – made by mixing cement, sand and water
- concrete – made by mixing cement, sand, rock chippings and water
- glass – made by heating up a mixture of limestone, sand and soda until it melts.

Reinforced concrete is a composite material made by allowing concrete to set around steel supports.

## Clues in Names

The name calcium carbon**ate** tells us that the compound contains calcium, carbon and lots of oxygen.

The names of some other compounds, such as potassium nitr**ite**, have a slightly different ending.

If a compound ends in 'ite', it contains some oxygen but not as much as if the name of the compound ended in 'ate'.

Potassium nitrite, therefore, must contain potassium, nitrogen and some oxygen.

## Making Salts

Salts are very important compounds. They can be made by reacting acids with metals, metal carbonates, metal oxides or metal hydroxides.

Sulfuric acid forms sulfate salts, while hydrochloric acid forms chloride salts.

## ther Rocks

- Granite is an example of an igneous rock. Igneous rocks are formed when molten rocks cool down and solidify.
- If the igneous rock formed quickly, it will contain small crystals, for example basalt.
- If the igneous rock formed slowly, it will contain large crystals, for example gabbro.
- If limestone is subjected to high temperatures or pressures, it can be turned into the metamorphic rock, marble.

**Marble**

### QUICK TEST

1. Name the main chemical compound in limestone.

2. What type of rock is limestone?

3. Name the products of the thermal decomposition of limestone.

4. By what name is calcium oxide also known?

5. By what name is calcium hydroxide also known?

### EXAM PRACTICE

1. Limestone is extracted from large quarries. The limestone from Hilltop Quarry has many uses.

| Use | Percentage of limestone from Hilltop Quarry (%) |
|---|---|
| To make calcium hydroxide | 40 |
| Construction | 30 |
| To make cement | 25 |
| To make mortar | 5 |

a) Compare the amount of limestone used to make mortar with the amount of limestone used to make calcium hydroxide.
   **(2 marks)**

b) Suggest why people living near to a proposed new limestone quarry might be concerned.   **(1 mark)**

## SUMMARY

- **Limestone contains calcium carbonate.**

- **The thermal decomposition of calcium carbonate produces calcium oxide and carbon dioxide.**

- **Limestone can be used to make calcium hydroxide, cement, mortar, concrete and glass.**

- **Igneous rocks are formed when molten rocks solidify; metamorphic rocks are formed when high temperatures and pressures change existing rocks.**

# Structure of the Earth

## The Layered Structure

The Earth has a layered structure.

- At the centre of the Earth is the **core**. The core is divided into two parts: the solid inner core and the outer core, which is liquid.
- The core is surrounded by the **mantle**.
- Around the mantle is the **crust**.

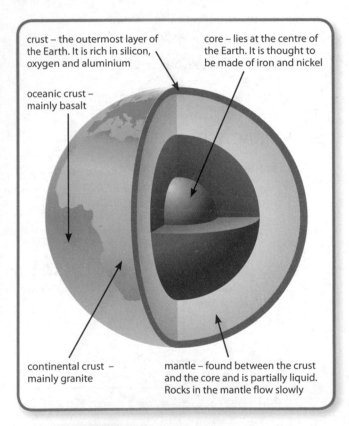

crust – the outermost layer of the Earth. It is rich in silicon, oxygen and aluminium

core – lies at the centre of the Earth. It is thought to be made of iron and nickel

oceanic crust – mainly basalt

continental crust – mainly granite

mantle – found between the crust and the core and is partially liquid. Rocks in the mantle flow slowly

## Plate Tectonics

People used to believe that the features that we see at the Earth's surface were formed as the Earth cooled down. Scientists now believe that these features were caused by **plate** tectonics.

In this theory, the Earth's **lithosphere** (crust and the upper part of the mantle) is broken into about a dozen pieces called plates. These plates are carried by convection currents in the Earth's mantle. The currents are caused by the heat released by natural radioactive decay. The plates move a few centimetres each year. This is about the same rate as your finger nails grow.

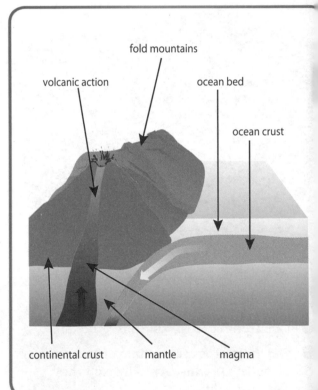

fold mountains

volcanic action

ocean bed

ocean crust

continental crust     mantle     magma

## Earthquakes

Sometimes the plates cannot move smoothly. **Earthquakes** occur where the plates try to move past each other but become stuck. The forces on these plates gradually build up. Eventually the plates move and they release the strain that has built up as an earthquake, or if it happens under water, a **tsunami**.

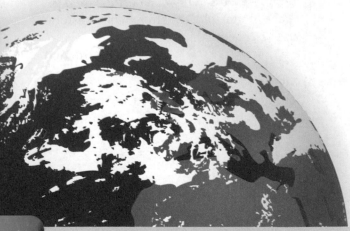

## Volcanoes

**Volcanoes** also occur at plate boundaries. When an oceanic plate collides with a continental plate, the denser oceanic plate is subducted beneath the continental plate. Some of the oceanic plate may melt to form magma (molten rock that is underground). The hot magma may be less dense than the surrounding rock; if it is, it may rise to the surface through cracks to form volcanoes. Volcanoes can cause devastating loss of life.

If the magma is iron-rich, it will be quite runny and the volcano will erupt relatively slowly and safely.

If the magma is silica-rich, however, it will be much more viscous. The volcano will erupt explosively, producing volcanic ash and throwing out molten rock called 'bombs'. These volcanoes are very dangerous to local people.

Although our methods of predicting when earthquakes and volcanoes will happen are improving, we still cannot say exactly when or where they will occur.

## Evidence for the Structure of the Earth

Evidence for the layered structure of the Earth comes from studies of seismic waves. These waves are sent out by earthquakes. The material through which the seismic wave travels affects the speed of the wave. This allows scientists to work out which parts of the Earth are solid and which are liquid.

## SUMMARY

- The Earth has a layered structure.

- The Earth's plates move slowly across the Earth's surface, causing earthquakes, tsunamis and volcanoes.

- Silica-rich volcanoes are especially dangerous.

- Our evidence for the structure of the Earth comes from our studies of seismic waves.

## QUESTIONS

### QUICK TEST

1. In what state is the inner core?

2. In what state is the outer core?

3. What is the name of the layer between the outer core and the crust?

4. Roughly how many plates are there?

5. What can form if an earthquake happens under the sea?

### EXAM PRACTICE

1. The Earth's lithosphere is split into about a dozen plates.

   a) What is the lithosphere? **(1 mark)**

   b) At what rate do these plates move? **(1 mark)**

   c) Before we knew about plate tectonics, how did people believe that geological features such as mountains were formed? **(1 mark)**

# Cosmetics

## Nail Varnish

A **solution** is a mixture made when a **solvent** dissolves a **solute**.

Water is a good solvent for many, but not all, substances. Some solutes, like nail varnish, are insoluble in water.

## Perfumes

Ethanol is often used as a solvent in perfumes. Traditional perfumes contain plant and animal extracts such as jasmine and musk, but these ingredients can be very expensive.

Today we often use cheaper, man-made fragrances such as esters. Esters are made by reacting carboxylic acids with alcohols.

A good perfume should:

- evaporate easily from the skin
- be non-toxic
- not react with water (sweat)
- not irritate the skin
- be insoluble in water so it is not washed off easily.

Perfumes evaporate because although there are strong forces of attraction **within** perfume molecules there are much weaker forces of attraction **between** perfume molecules.

When perfume is put on the skin, some of the molecules gain enough heat energy to evaporate.

## Animal Testing

Cosmetic products have to be tested before they can be sold. Some tests involve living animals. Some people believe that this causes avoidable suffering to animals while other people think that animal testing is the best way to ensure that products are safe for people to use.

# Making Ethanol

**Ethanol** is a type of alcohol. It has the following structure:

$$H-\overset{\displaystyle H}{\underset{\displaystyle H}{C}}-\overset{\displaystyle H}{\underset{\displaystyle H}{C}}-O-H$$

Ethanol can be made from fermenting sugar cane or sugar beet.

Ethanol is a useful biofuel (a fuel made from living materials). Unlike petrol, ethanol is a renewable fuel, but large areas of fertile land must be used to grow the sugar cane or sugar beet.

## Fermentation

Ethanol is produced by **fermentation**. During fermentation, yeast converts glucose (sugar) into ethanol and carbon dioxide:

$$\text{glucose} \xrightarrow{\textit{yeast}} \text{ethanol} + \text{carbon dioxide}$$

$$C_6H_{12}O_6 \rightarrow 2C_2H_5OH + 2CO_2$$

Fermentation has been used to produce alcoholic drinks such as wine and beer. The consumption of alcoholic drinks causes many social and health problems and is banned by some religions.

In some countries the sugar from sugar cane or sugar beet is used to make alcohol. The alcohol can be mixed with petrol to make a fuel for vehicles. Ethanol burns very cleanly and it is a renewable resource.

## SUMMARY

- Ethanol is used as a solvent in perfumes.
- Cosmetic products are tested on animals but many people object to the idea of animal testing.
- Ethanol can be produced by fermentation.
- Ethanol produced by fermentation can be used as a renewable fuel.

## QUESTIONS

### QUICK TEST

1. What is a solution?

2. Name a solute that is insoluble in water.

3. Name a solvent often used in perfumes.

4. Write down a plant extract used in perfumes.

5. How can an ester be made?

### EXAM PRACTICE

1. Ethanol is a type of alcohol.

   a) What is the name of the reaction in which glucose is converted into alcohol and carbon dioxide? **(1 mark)**

   b) Write a word equation for this reaction. **(1 mark)**

# Chemicals

## Uses

It is very useful to be able to identify chemicals.

Different chemicals have different uses.

| Chemical | Use |
|----------|-----|
| Ammonia | To make fertilisers |
| Carbohydrate | A type of food we need for energy |
| Carbon dioxide | To make carbonated drinks |
| Caustic soda | To make bleach / oven cleaner |
| Citric acid | To flavour food and drink |
| Ethanoic (acetic) acid | In vinegar |
| Hydrochloric acid | To remove limescale from boilers |
| Phosphoric acid | A catalyst for the production of industrial alcohol |
| Sodium chloride | To 'grit' roads in winter |
| Water | In industry and in agriculture |

## Gas Tests

### Hydrogen

Burns with a 'squeaky' pop.

### Oxygen

Relights a glowing splint.

### Carbon dioxide

When bubbled through limewater, it turns the limewater cloudy.

### Ammonia

Turns damp red litmus paper blue.

### Chlorine

Bleaches damp litmus paper.

## Collecting Gases

- Upward delivery is used to collect gases that are less dense than air.
- Downward delivery is used to collect gases that are denser than air.
- Gases that are fairly insoluble are collected over water.
- If we need to measure the volume of gas, we can collect it using a gas syringe.

## Hazard Symbols

**Hazard symbols** warn us about chemicals.

| Oxidising | Highly flammable |
|-----------|------------------|
| Allows other materials to burn more fiercely | Catches fire easily |
| **Toxic** | **Harmful** |
| Can cause death | Less dangerous than toxic |
| **Corrosive** | **Irritant** |
| Attacks and destroys living tissue | Can cause reddening or blistering of skin |

# Flame Tests

Flame tests can be used to identify the metals in compounds. The colour of the flame indicates the metal present:

- lithium – red
- sodium – orange/ yellow
- potassium – lilac
- calcium – brick red
- barium – apple green
- copper – blue/green.

# Hydroxide Tests

We can identify the metals present in metal salt solutions by adding **sodium hydroxide solution**.

If the metal ion forms a **precipitate**, we can use the colour of the precipitate to identify the metal present:

- copper(II) – pale blue precipitate
- iron(II) – green precipitate
- iron(III) – brown precipitate.

# Other Tests

| Ion | What to do | What happens |
|---|---|---|
| Ammonium | React with sodium hydroxide, then test gas produced (ammonia) | Damp red litmus paper goes blue |
| Carbonate | React with dilute acid, then bubble the gas that is produced through limewater (carbon dioxide) | Limewater goes cloudy |
| Sulfate | Add dilute hydrochloric acid, then barium chloride solution | A white precipitate forms |

Modern instrumental methods are accurate, sensitive and quick.

## SUMMARY

- Chemicals have many different uses.
- Scientists use a range of tests to identify common chemicals.
- Hazard symbols are used to warn people about dangerous chemicals.

## QUESTIONS

### QUICK TEST

1. Which gas burns with a 'squeaky' pop?

2. What type of chemical allows other chemicals to burn more easily?

3. What type of chemical can cause death?

4. What does a corrosive chemical do?

5. What colour would you expect when a sample of a sodium salt is used in a flame test?

### EXAM PRACTICE

1. Use the words below to complete the sentences.

   **carbon dioxide**        **ammonia**

   **chlorine**        **oxygen**

   If ____**a)**____ is bubbled through limewater, the limewater turns cloudy.

   ____**b)**____ relights a glowing splint.

   ____**c)**____ bleaches damp litmus paper.

   ____**d)**____ turns damp red litmus paper blue.

   **(4 marks)**

# Answers

## Answering Quality of Written Communication Questions

A number of the questions in your examinations will include an assessment of the quality of your written communication (QWC). These questions are worth a maximum of 6 marks.

Your answers to these questions will be marked according to...
- the level of your understanding of the relevant science
- how well you structure your answer
- the style of your writing, including the quality of your punctuation, grammar and spelling.

- To achieve 5–6 marks, your answer should contain relevant science, and be organised and presented in a structured and coherent manner. You should use scientific terms appropriately and your spelling, punctuation and grammar should have very few errors.
- To achieve 3–4 marks, there may be some errors in your spelling, punctuation and grammar, and your answer will miss some of the things expected.
- To achieve 1–2 marks, your answer will be written using simplistic language. You will have included some relevant science, but the quality of your written communication may have limited how well the examiner can understand your answer. This could be due to lots of errors in spelling, punctuation and grammar, misuse of scientific terms or a poor structure.

The answers to the QWC questions in this guide are provided as a list of key points and then as a model answer.

## Day 1

*pages 4–5*
### How Science Works
QUICK TEST
1. The variable we choose to change in an experiment
2. The variable that we measure in an experiment
3. A variable that can be put in order
4. A variable that can have any whole-number value
5. It is close to the true value
6. When the independent variable and the dependent variable are both continuous.

EXAM PRACTICE
1. **a)** The force applied **(1)**
   **b)** The length of the spring **(1)**
   **c)** **Accept any suitable answer, e.g.** Using the same spring **(1)**

*pages 6–7*
### Atomic Structure
QUICK TEST
1. They have the same number of protons
2. Nucleus
3. Electrons
4. In the periodic table
5. Approx. 100

EXAM PRACTICE
1. **a)** calcium carbonate + hydrochloric acid → calcium chloride + water + carbon dioxide **(1)**

**b)** One carbon atom and two oxygen atoms **(1)**

*pages 8–9*
### The Periodic Table
QUICK TEST
1. Groups
2. Periods
3. The number of protons plus the number of neutrons
4. Atomic number / proton number / number of protons
5. Mass number / number of neutrons

EXAM PRACTICE
1. 6 protons, 6 electrons and 7 neutrons **(1)**

*pages 10–11*
### Electronic Structure
QUICK TEST
1. Shells / levels around the nucleus of an atom
2. Two
3. 2,1
4. Group 1
5. Group 4

EXAM PRACTICE
1. **a)** 2,8,2 **(1)**     **b)** Group 2 **(1)**

*pages 12–13*
### Ionic Bonding
QUICK TEST
1. An atom or group of atoms with a charge

2. 1–
3. 1+
4. 2–
5. 2+

EXAM PRACTICE
1. **a)** It has lots of strong ionic bonds **(1)** that require a lot energy to separate them **(1**
   **b)** sodium + chlorine → sodium chloride **(1)**
   **c)** 1+ **(1)**

*pages 14–15*
### Covalent Bonding
QUICK TEST
1. A shared pair of electrons
2. There is an electrostatic attraction between the nuclei of the atoms and the bonding electrons
3. $CH_4$
4. $NH_3$
5. Silicon dioxide

EXAM PRACTICE
1. Substance A is a simple covalen structure **(1)**. It has a relatively low boiling point because there are only weak forces of attraction between molecules **(1)**. It does not conduct electricity because it does not contain ions or free electrons **(**

## kali Metals
On the far left-hand side / In Group 1
Potassium
Hydrogen
Orange
Lilac

**AM PRACTICE**

**a)** sodium + water → sodium hydroxide + hydrogen **(1)**

**b)** $2Na + 2H_2O \rightarrow 2NaOH + H_2$ **(1 mark for correct elements, 1 mark for balancing equation)**

**c)** So it does not react with air or water **(1)**

**d)** The outer electron is further from the nucleus **(1)** so it is lost more easily **(1)**

## ay 2
## ble Gases and Halogens
Halogens / Group 7
Noble gases / Group 0
Liquid
Balloons
Filament lamps

**AM PRACTICE**

Chlorine is more reactive than iodine **(1)** because in this reaction chlorine displaces iodine from a solution of potassium iodide **(1)**.

## lculations 1
To compare the masses of different atoms
Carbon-12
The relative formula mass of a substance in grams
It does not go to completion

---

5. The percentage yield of a reaction $= \dfrac{\text{actual amount of product}}{\text{theoretical yield}} \times 100\%$

**EXAM PRACTICE**

**1. a) Any one from:** The reaction is reversible and does not go to completion; Some of the product is lost, for example during filtering or evaporation; There may be side reactions that are producing another product **(1)**

**b)** 88% **(1)**

*pages 22–23*
## Calculations 2
QUICK TEST

**1. a)** 0.05 moles
**b)** 0.2 moles
**c)** 0.25 moles
**2. a)** 0.0025 moles
**b)** 0.01 moles

EXAM PRACTICE

**1.** Phenolphthalein **(1)**

*pages 24–25*
## Calculations 3
QUICK TEST

**1.** The simplest whole number ratio of atoms in the formula of a substance
**2.** $CH_2$
**3.** $NH_2$
**4.** $kJ\,mol^{-1}$
**5.** Exothermic

EXAM PRACTICE

**1.** The reaction is exothermic **(1)**, overall more energy is given out than is taken in **(1)**

*pages 26–27*
## The Haber Process
QUICK TEST

**1.** Fractional distillation of liquid air
**2.** Natural gas
**3.** Iron
**4.** Around 450 °C
**5.** Around 200 atmospheres

---

EXAM PRACTICE

**1. Key points:**

- $N_2 + 3H_2 \rightleftharpoons 2NH_3$
- An iron catalyst is used.
- This increases the rate of reaction.
- Pressure of 200 atmospheres used.
- This increases the yield of ammonia.
- This increases the rate of reaction.
- A moderate temperature of around 450°C is used.
- This gives a reasonable yield of ammonia and a reasonable rate of reaction.

**Model answer:**

The symbol equation is $N_2 + 3H_2 \rightleftharpoons 2NH_3$. Ammonia is made using an iron catalyst, which increases the rate of reaction. A pressure of 200 atmospheres is used, which increases the yield of ammonia. A moderate temperature of around 450°C is used, which gives a reasonable yield of ammonia and a reasonable rate of reaction.

*pages 28–29*
## The Manufacture of Sulfuric Acid
QUICK TEST

**1.** Where nothing can enter or leave
**2.** When the rate of forward reaction = rate of reverse reaction
**3.** Vanadium(v) oxide
**4.** 450 °C
**5.** 2 atmospheres

EXAM PRACTICE

**1. a)** Sulfur is burnt in oxygen **(1)**
**b)** It is a reversible reaction **(1)**
**c)** It increases the rate of reaction **(1)** but has no effect on the position of equilibrium **(1)**

# Answers

## Day 3

pages 30–31
### Rates of Reaction
QUICK TEST
1. They need to collide, and when they do collide they must have enough energy to react (activation energy)
2. It increases the rate of reaction
3. It decreases the rate of reaction
4. It increases the rate of reaction
5. At the start, when there is a higher concentration of reactants

EXAM PRACTICE
1. **Key points:**
   **Temperature**
   • If the temperature was increased the rate of reaction would increase.
   • Particles move faster
   • Particles have more energy.
   • Particles collide more often.
   • More have enough energy to react
   • Particles have the activation energy.
   **Concentration**
   • If the concentration of the hydrochloric acid was increased the rate of reaction increases.
   • The reactant particles are closer together.
   • Particles collide more often.
   **Model answer:**
   If the temperature increases, the rate of reaction increases. The particles move faster and collide more often. More particles have enough energy to react (i.e. they have activation energy).
   If the concentration of the hydrochloric acid increases, the rate of reaction increases. The reactant particles are now closer together (i.e. they collide more often).

pages 32–33
### Energy
QUICK TEST
1. Exothermic
2. Endothermic
3. Energy
4. A reversible reaction
5. Endothermic

EXAM PRACTICE
1. a) hydrated copper sulfate $\rightleftharpoons$ **(1)** anhydrous copper sulfate + water **(1)**
   b) Changes colour from blue to white **(1)**
   c) Add it to white anhydrous copper sulfate and it would turn blue **(1)**

pages 34–35
### Calorimetry
QUICK TEST
1. Calorimetry
2. joules (J)
3. °C
4. grams (g)
5. 4.2 joules

EXAM PRACTICE
1. a) Exothermic reaction **(1)**
   b) $5 \times 4.2 \times 4$ **(1)** $= 84$ J **(1)**

pages 36–37
### Energy Profile Diagrams
QUICK TEST
1. Energy
2. The minimum amount of energy needed to break the old bonds
3. Energy
4. Increase the rate of a reaction
5. They are not used up during a reaction

EXAM PRACTICE
1. a) Energy profile diagram drawn with the reactants higher in energy than the products **(1)**; Both axes labelled **(1)**; Activation energy labelled **(1)**; Energy given out by the reaction labelled **(1)**

b) The iron catalyst works by offering an alternative reaction pathway with lower activation energy **(1)**

pages 38–39
### Aluminium and Dilute Sulfuric Acid
QUICK TEST
1. Aluminium oxide
2. Cryolite
3. No
4. Yes
5. Aluminium

EXAM PRACTICE
1. a) $Al^{3+} + 3e^- \rightarrow Al$ **(1)**
   b) $2O^{2-} \rightarrow O_2 + 4e^-$ **(1)**
   c) The oxygen that is produced reacts to form carbon dioxide which erodes the electrode (

pages 40–41
### Sodium Chloride
QUICK TEST
1. **Accept any suitable answer, e.g.** The production of fertilise As colouring agents; In firewor
2. Barium sulfate
3. barium chloride + sodium sulfate $\rightarrow$ barium sulfate + sodium chloride
4. Solid
5. Aqueous / dissolved in water

EXAM PRACTICE
1. When it is solid the ions cannot move so the lead bromide cann conduct electricity **(1)**. When it i molten the ions can move **(1)**.
   At the anode $2Br^- \rightarrow Br_2 + 2e^-$ (
   At the cathode $Pb^{2+} + 2e^- \rightarrow Pb$ (

pages 42–43
### Copper
QUICK TEST
1. By heating
2. By electrolysis
3. Anode
4. Cathode
5. $Cu^{2+}$

a) $Cu(s) \rightarrow Cu^{2+}(aq) + 2e^-$
   **(1 mark for equation, 1 for state symbols)**

b) $Cu^{2+}(aq) + 2e^- \rightarrow Cu(s)$
   **(1 mark for equation, 1 for state symbols)**

## Day 4

*pages 44–45*
### Acids and Alkalis
QUICK TEST
1. It is only partially ionised in water
2. $OH^-$
3. Hydrochloric acid is a strong acid and is completely ionised in water
4. **Accept any suitable answer, e.g.** hydrochloric acid, sulfuric acid and nitric acid
5. **Accept any suitable answer, e.g.** Ammonia

EXAM PRACTICE
1. A gas / carbon dioxide was produced in both reactions **(1)**. Sulfuric acid is a strong acid, so has a higher concentration of hydrogen ions than ethanoic acid, which is a weak acid **(1)**. This means that the hydrogen ions in sulfuric acid will collide more quickly with calcium carbonate particles, so the rate of reaction will be higher. **(1)**

*pages 46–47*
### Making Salts
QUICK TEST
1. Neutral
2. Alkali
3. Acid
4. Magnesium chloride
5. Calcium chloride

EXAM PRACTICE
1. a) calcium carbonate + hydrochloric acid → calcium chloride + water + carbon dioxide **(1)**
   b) **Accept:** 1–3 **(1)**

*pages 48–49*
### Metals
QUICK TEST
1. It is the attraction between positive metal ions and negative delocalised electrons
2. Alloys
3. Iron
4. **Accept any suitable answer, e.g.** It turns limewater cloudy

EXAM PRACTICE
1. **Key points:**
   **Metallic structure consists of:**
   • Positive metal ions.
   • Surrounded by delocalised electrons.
   **Melting points:**
   • Metals have high melting points
   • Metals are solid at room temperature (apart from mercury).
   • Strong attraction between the positive metals ions and the delocalised electrons.
   **Electrical conductivity:**
   • Metals are good electrical conductors.
   • The delocalised electrons can move throughout the structure.

**Model answer:**
A metallic structure consists of positive metal ions surrounded by delocalised electrons. This structure means that there is a strong attraction between the positive metal ions and the delocalised electrons. Metals have high melting points and are solid at room temperature (with the exception of mercury).The delocalised electrons can move throughout the structure, so metals are good conductors of electricity.

*pages 50–51*
### Useful Metals
QUICK TEST
1. It is too soft

2. Alloy
3. Copper and zinc
4. **Accept any suitable answer, e.g.** Drinks cans; Bicycles; Aeroplanes
5. Rutile

EXAM PRACTICE
1. **Key points:**
   **Extraction:**
   • By electrolysis
   • Of bauxite
   **Advantages**
   • Aluminium bodies are lighter.
   • Aluminium cars have better fuel economy.
   • Aluminium does not corrode
   • Aluminium lasts longer.
   **Disadvantage**
   • Aluminium is more expensive to produce.

**Model answer:**
Aluminium is extracted from its ore by the electrolysis of bauxite. There are several advantages of using aluminium to make a car body. Aluminium is light, so aluminium cars have a better fuel economy than a heavier steel car. Aluminium doesn't corrode, so aluminium car bodies last longer than steel car bodies. However, aluminium is expensive to extract, so cars made from aluminium are more expensive to produce and buy.

*pages 52–53*
### Iron and Steel
QUICK TEST
1. Iron(III) oxide
2. Hard but brittle
3. In a blast furnace
4. It has reduced
5. 96%

EXAM PRACTICE
1. a) wrought iron **(1)**
   b) low carbon steel **(1)**
   c) cast iron **(1)**
   d) iron(III) oxide **(1)**

# Answers

*pages 54–55*
**Water**

QUICK TEST
1. A rock that contains water
2. To remove suspended particles like clay
3. To reduce the number of microorganisms to acceptable levels
4. It uses a lot of energy
5. Nitrates / Phosphates

EXAM PRACTICE
1. **Key points:**
   • Add nitric acid.
   • Then add silver nitrate solution.
   • A white precipitate indicates chloride ions.
   • A cream precipitate indicates bromide ions.
   • A yellow precipitate indicates iodide ions.
   • The compounds formed are silver chloride, silver bromide and silver iodide.

**Model answer:**
Chloride, bromide and iodide ions can be detected in water samples by first adding nitric acid, and then adding silver nitrate solution. The formation of a white precipitate indicates chloride ions, a cream precipitate indicates bromide ions, and a yellow precipitate indicates iodide ions. The compounds formed are silver chloride, silver bromide and silver iodide.

## Day 5

*pages 56–57*
**Hard and Soft Water**

QUICK TEST
1. The Sun warms the sea
2. Carbon dioxide
3. Increases solubility
4. Decreases solubility
5. Increases solubility

EXAM PRACTICE
1. Increasing the temperature decreases the amount of carbon dioxide that is dissolved in the drink **(1)**

*pages 58–59*
**Detergents**

QUICK TEST
1. To remove coloured stains
2. To remove stains at low temperatures
3. To make them more attractive
4. It helps the water to drain away from the crockery / cutlery
5. The maximum water temperature to use

EXAM PRACTICE
1. **a)** To remove the dirt from the fabric **(1)**
   **b)** An acid **(1)**
   **c)** Hydrophilic end **(1)**

*pages 60–61*
**Special Materials**

QUICK TEST
1. **Accept any suitable answer, e.g.** Pencil lead; Lubricant; Electrodes
2. $C_{60}$
3. Fullerene, diamond and graphite
4. Lustrous and colourless
5. Black

EXAM PRACTICE
1. **a)** Each carbon atom is bonded to four others **(1)** by strong covalent bonds **(1)**
   **b)** Lots of strong **(1)** covalent bonds **(1)**
   **c)** There are no free electrons / ions to move **(1)**

*pages 62–63*
**New Materials**

QUICK TEST
1. It is too big
2. Nylon
3. Teflon
4. To make non-stick saucepans
5. Bulletproof vests

EXAM PRACTICE
1. **a)** It contains very small fibres
   **b)** The small fibres trap air **(1)**. This acts as a layer of insulation **(1)** which stops body heat from being lost **(1**

*pages 64–65*
**Vegetable Oils**

QUICK TEST
1. A and D
2. Sunflower oil and olive oil
3. Seeds, nuts and fruits
4. They have carbon double bond
5. Vinegar and vegetable oil

EXAM PRACTICE
1. **a)** emulsion **(1)**
   **b)** polyunsaturated **(1)**
   **c)** biofuel **(1)**
   **d)** bromine water **(1)**

*pages 66–67*
**Food Additives**

QUICK TEST
1. Throughout the European Union
2. Keeps unblendable liquids mixed together
3. Natural
4. Colourings
5. Antioxidant

EXAM PRACTICE
1. Sodium hydrogencarbonate **(1**
2. **Accept any suitable answer, e.g.** To make the dough rise; To improve the texture **(1)**
3. carbon dioxide **(1)**

*pages 68–69*
**Crude Oil**

QUICK TEST
1. A compound that only contain carbon and hydrogen atoms
2. A part of crude oil / a group of molecules with a similar numb of carbon atoms
3. At the top
4. Bitumen
5. Kerosene

**a)** fractional distillation **(1)**
**b)** evaporates **(1)**
**c)** condense **(1)**
**d)** fractions **(1)**

## Day 6

### pages 70–71
### Fuels
QUICK TEST
1. Hydrogen and carbon
2. Carbon dioxide and water vapour
3. Carbon
4. Global warming / greenhouse effect
5. It produces soot / carbon monoxide
EXAM PRACTICE
1. carbon dioxide **(1)**
2. Carbon / soot / carbon monoxide **(1)**
3. water (vapour) **(1)**

### pages 72–73
### Air and Air Pollution
QUICK TEST
1. Nitrogen
2. Carbon dioxide
3. Plants evolved
4. Oxygen
5. Sulfur dioxide
EXAM PRACTICE
1. **a)** Mars / Venus **(1)**
   **b)** Plants evolved **(1)** and carry out photosynthesis **(1)**
   **c)** Global warming / Greenhouse effect **(1)**

### pages 74–75
### Pollution
QUICK TEST
1. Creates new jobs
2. Noise / dust / lorries
3. Plastic
4. It can produce toxic gases
5. It is hard to separate them out
EXAM PRACTICE
1. **a)** It can absorb water **(1)**

   **b)** It is difficult for bacteria / mould to grow **(1)** so food stays fresh for longer **(1)**

### pages 76–77
### Alkanes and Alkenes
QUICK TEST
1. $C_nH_{2n+2}$
2. Fuel
3. (Covalent) bonds
4. Cracking
5. $C_2H_4$
EXAM PRACTICE
1. **a)** Alkenes **(1)**
   **b)** It has carbon double bonds **(1)**
   **c)** ethene + steam → ethanol **(1)**

### pages 78–79
### Alcohols, Acids and Esters
QUICK TEST
1. Alcohols
2. Esters
3. Alcohols
4. Carboxylic acids
5. Any number under 7
EXAM PRACTICE
1. **a)** 7 **(1)**
   **b)** **Accept one suitable answer, e.g.** Solvent; Fuel; Drinks **(1)**
   **c)** Ethanoic acid **(1)**

## Day 7

### pages 80–81
### Polymers
QUICK TEST
1. Monomers
2. Polyethene (polythene)
3. Polypropene
4. It makes it more flexible
5. It will not soften but may eventually burn
EXAM PRACTICE
1. Ethene **(1)**
2. Plastic bags / bottles **(1)**
3. It increases the melting point **(1)** because there are now greater forces of attraction between the polymer chains. **(1)**

### pages 82–83
### Limestone
QUICK TEST
1. Calcium carbonate
2. Sedimentary
3. Calcium oxide (quicklime) and carbon dioxide
4. Quicklime
5. Slaked lime
EXAM PRACTICE
1. **a)** Eight times as much limestone is used to make calcium hydroxide as is used to make mortar **(2)** **(For 1 mark: 35% more limestone is used to make calcium hydroxide than is used to make mortar)**
   **b)** **Accept one from:** Noise; Dust; Damage to the environment **(1)**

### pages 84–85
### Structure of the Earth
QUICK TEST
1. Solid
2. Liquid
3. Mantle
4. Dozen / twelve
5. Tsunami
EXAM PRACTICE
1. **a)** The crust and upper mantle **(1)**
   **b)** A few centimetres per year **(1)**
   **c)** Mountains formed as the Earth cooled down **(1)**

### pages 86–87
### Cosmetics
QUICK TEST
1. A mixture made when a solute dissolves in a solvent
2. **Accept any suitable answer, e.g.** Nail varnish
3. Ethanol
4. **Accept any suitable answer, e.g.** Jasmine; Lavender
5. By reacting an alcohol with a carboxylic acid

EXAM PRACTICE
1. **a)** Fermentation **(1)**
   **b)** glucose → ethanol + carbon dioxide **(1)**

*pages 88–89*
## Chemicals
QUICK TEST
1. Hydrogen
2. Oxidising chemical
3. Toxic chemical / Radioactive chemical
4. Attacks and destroys living tissue
5. Orange / yellow

EXAM PRACTICE
1. **a)** carbon dioxide **(1)**
   **b)** oxygen **(1)**
   **c)** chlorine **(1)**
   **d)** ammonia **(1)**

| LS | 6.14 |
|----|------|
|    |      |
|    |      |
|    |      |
|    |      |
|    |      |
|    |      |
|    |      |